THE END OF
BRITISH FARMING

ANDREW O'HAGAN

D0756165

P

PROFILE BOOKS
and the
LONDON REVIEW OF BOOKS

First published in Great Britain in 2001 by
Profile Books Ltd
58A Hatton Garden
London EC1N 8LX
www.profilebooks.co.uk

with

London Review of Books
28–30 Little Russell Street
London WC1A 2HN
www.lrb.co.uk

1 3 5 7 9 10 8 6 4 2

Typeset in Quadraat
Designed by Peter Campbell
Printed and bound in Great Britain by
Bookmarque Ltd, Croydon, Surrey

ISBN 1 86197 392 6

Now and again a finch, a starling or a sparrow would come meaning to drink – athirst from the meadow or the cornfield – and start and almost entangle their wings in the bushes, so completely astonished that anyone should be there.

Richard Jefferies, *The Life of the Fields*, 1884.

THE END OF BRITISH FARMING

This last while I have carried my heart in my boots. For a minute or two I actually imagined I could be responsible for the spread of foot and mouth disease across Britain. On my first acquaintance with the hill farmers of the Lake District, on a plot high above Keswick, I had a view of the countryside for tens of miles. I thought of the fields that had passed underfoot, all the way back to Essex, through Dumfriesshire, Northumberland or Sussex. Later I would continue on my way to Devon, passing through other places waking up in the middle of the worst agricultural nightmare in seventy years. My boots are without guilt, but in all the walking here and there, in the asking and listening, I came to feel that British farming was already dying, that the new epidemic was but an unexpected acceleration of a certain decline.

In the last two months nearly a million head of livestock have been condemned. The industry has lost hundreds of millions of pounds. A freeze still holds on the export of livestock. Country footpaths are zones of reproach and supermarkets are running out of Argentinian beef. The Agriculture Minister, Nick Brown, is accused of doing too much and doing too little. The questions surrounding the foot and mouth epidemic – where will it all end? how did it all start? – might be understood to accord with anxiety about every aspect of British agriculture today. The worst has not been and gone. It is yet to come. Still, one thing may already be clear: British farming hanged itself on the expectation of plenty.

One day not long ago I was in the Sainsbury's superstore on the Cromwell Road. Three of the company's top brass ushered me down the aisles, pointing here, gasping there, each of them in something of a swoon at the heavenliness on offer. 'People want to be interested,' said Alison Austin, a technical adviser, 'you've just got to capture their imagination.' We were standing by the sandwiches and the takeaway hot foods lined up in front of the whooshing doors. Alison swept her hand over the colourful bazaar of sandwich choices. 'This is a range called Be Good to Yourself,' she said, 'with fresh, healthy fillings, and here we have the more gourmet range, Taste the Difference. We have a policy of using British produce where we can. With carrots, for example, we want to provide economic profitability to the farmer, using the short carrots for one line of produce and the bigger ones for another.'

The Cromwell Road branch of Sainsbury's is what they call a 'flagship store'. It's not only a giant emporium, it is also grander than any other store in the chain, selling more champagne, fresh fish, organic meat and Special Selection food. Six varieties of caviar are available all year round.

'People are gaining more confidence in sushi,' said Peter Morrison, Manager, Trading Division. 'We have joined forces with very credible traders such as Yo! Sushi and we aim to educate customers by bringing them here.' Alison handed me a cup of liquid grass from the fresh juice bar, Crussh. There was something unusually potent about that afternoon – the thoughts in my head as I tilted the cup – and for a moment the whole supermarket seemed to spin around me. People wandered by. The place was a madhouse of bleeping barcodes. 'How do you like it?' one of them asked. I gulped it down and focused my eyes. 'It tastes like an English field,' I said.

The store manager guided me to the cut flowers. 'We are the UK's largest flower sellers,' he told me, 'the biggest year on year increase of any product in the store is in flowers.' The bunches before me were a far cry from the sad carnations and petrol station bouquets that now lie about the country as tributes to the suddenly dead. The ones he showed me had a very smart, sculptural appearance, and they sold for £25 a pop. 'We have 40 kinds of apple,' Alison said, 'and again, we take the crop, the smaller ones being more for the economy bags.'

'Someone came in on Christmas Eve and asked for banana leaves,' the keen young product manager over in fruit and vegetables told me, 'and you know something? We had them.'

You would have to say that Sainsbury's is amazing. It has everything – 50 kinds of tea, 400 kinds of bread, kosher chicken schnitzels, Cornish pilchards – and everywhere I turned that day there was some bamboozling elixir of the notion of plenty. Their own-brand products are made to high standards: the fresh meat, for example, is subject to much higher vigilance over date and provenance than any meat in Europe.[1] 'Some things take a while,' Peter Morrison said, 'you can put something out and it won't work. Then you have to think again, about how to market it, how to package it, where to place it, and six months later you'll try again and it might work.' We stopped beside the yoghurts. 'Now this,' he said, picking up a tub of Devon yoghurt, 'is made at a place called Stapleton Farm. We got wind of how good it was: a tiny operation, we went down there, we got some technical advisers involved, and now look, it's brilliant!' I tasted some of the Stapleton yoghurt. It was much better than the

4

liquid grass. 'It's about the rural business growing,' Peter said. 'Real food is what people want. This couple in Devon' – he gestured to the yoghurt pots – 'started from virtually nowhere. Of course they were nervous at first about working with such a major retailer. But these people are the new kind of producer.'

Passing the condiments aisle I saw an old man standing in front of the Oxo cubes. He looked a bit shaky. His lips were moving and he had one of the foil-wrapped Oxo cubes in the palm of his hand. 'People go to Tuscany,' Alison was saying, 'and they eat Parma ham and they come back here and they want it all the time. So we go out and find the best.' You are always alone with the oddness of modern consumption. Walking under the white lights of Sainsbury's you find out just who you are. The reams of cartons, the pyramids of tins: there they stand on the miles of shelves, the story of how we live now. Cereal boxes look out at you with their breakfast-ready smiles, containing flakes of bran, handfuls of oats, which come from fields mentioned in the Domesday Book. And here you are in the year 2001 – choosing. We went over to the aisle with the cooking oils and Alison did one of her long arm-flourishes: 'When I was a child,' she said, 'my mother used a bottle of prescription olive oil to clean the salad bowl. Now look!' A line of tank-green bottles stretched into the distance. 'Choice!' she said.

Supermarket people like to use certain words. When you are with them in the fruit department they all say 'fresh' and 'juicy' and 'variety' and 'good farming practices'. (Or as head office puts it, 'in 1992 Sainsbury developed a protocol for growing crops under Integrated Crop Management System principles. Following these principles can result in reduced usage of pesticides by combining more traditional aspects of agriculture and new technologies.') In the meat department there is much talk of 'friendly', 'animal well-being', 'humane', 'safe', 'high standards' and 'provenance'. The executives spent their time with me highlighting what they see as the strength of the partnerships with British farming which keep everyone happy. 'The consumer is what matters,' said Alison, 'and we believe in strong, creative, ethical retailing.'

Down at the front of the store again I put one of the gourmet sandwiches on a table and opened it up. The bread was grainy. The lettuce was pale green and fresh. Pieces of chicken and strips of pepper were neatly set out on a thin layer of butter. The open sandwich was a tableau of unwritten biographies: grains and vegetables and meat were glistening there, uncontroversially, their stories of economic life and farming history and current disaster safely behind them.

When I was a boy we had a painting above the phone table. It was the only real painting in the house, and it showed a wide field in the evening with a farm at the far end. The farmhouse had a light in one of the windows. The painting had been a wedding present, and my mother thought it was a bit dour and dirty-looking, so she did the frame up with some white gloss, which flaked over the years. I used to lie on the hall carpet and look at the picture of the farm for ages; the field was golden enough to run through and get lost in, and the brown daubs of farmhouse were enough to send me into a swoon of God-knows-what. I suppose it was all part of a general childhood boredom, and it meant nothing, but it seemed very heightening at the time. The painting raised my feelings up on stilts, and made me imagine myself to be part of an older world, where people lived and worked in a state of sentimental peace. All rot of course. But lovely rot. Sometimes I would come downstairs in the night and shine my torch on the painting.

At one time it seemed as if all the farms around our way had been abandoned or pulled down to make room for housing. Past railway lines and beyond the diminishing fields we would find old, dilapidated Ayrshire farmhouses with rusted tractors and old wooden drinking troughs lying about in the yard, and we'd play in them for half the summer. Cranberry

Moss Farm, McLaughlin's Farm on Byrehill, Ashgrove Farm, the Old Mains – nowadays they are all buried under concrete, except for the farm at Toddhill, which became a home for the mentally handicapped. In my youth they had been like haunted houses. There were echoes in the barns.

Those farms seemed as remote from the daily reality of our lives as the one in the wedding picture. We would never live there: computer factories and industrial cleaners would soon replace them as providers of jobs, and it was these new places, in our Ayrshire, that spoke of the lives we were supposed one day to live. We took it for granted – much too early, as it turned out – that farming was a thing of the past, a thing people did before they were sophisticated like us. We never considered the stuff on our plates; we thought the school milk came on a lorry from London. Never for a second did my friends and I think of ourselves as coming from a rural community; like all British suburban kids, we lived as dark, twinkling fallout from a big city, in our case Glasgow; and we thought carports and breezeblocks were part of the natural order.

But of course there was plenty of agriculture. It surrounded us. The farms had just been pushed out a wee bit – and wee could seem larger than it was, at least for us, shocked by the whiteness of our new buildings into thinking a thatched roof was the height of exotic. Everything changed for me with the discovery of Robert Burns: those torn-up fields out there were his fields, those bulldozed farms as old as his words, both old and new to me then. Burns was ever a slave to the farming business: he is the patron saint of struggling farmers and poor soil. But in actual fact, despite our thoughts and our recovery from our thoughts, in the early 1970s British farming was in a pretty good state. J.G.S. and Frances Donaldson's *Farming in Britain Today*, published around this time, just before Britain's entry into the Common Market, expressed the view that a beautiful balance had been struck.

> Today agriculture is one of Britain's most efficient industries. It has a controlled growth of 3.5 per cent a year, and in the last ten years its labour productivity has increased at twice the rate for industry as a whole. It supplies approximately 50 per cent of the nation's food. Travelling through England today with its trim hedges, arable and ley farming, highly capitalised and intensively used buildings, it is hard to imagine the broken-backed appearance of yesterday.

'Yesterday' meant the 1920s and 1930s. But now, as I write, the situation of farming in this country is perhaps worse than it has ever been, and the countryside itself is dying. We are at a stage where it is difficult to imagine British farming surviving in any of its traditional forms; and for millions living on these islands, a long-term crisis has been turning into a terminal disaster.

Three years ago agriculture contributed £6.9 billion to the British economy, around 1 per cent of the Gross Domestic Product. It represented 5.3 per cent of the value of UK exports. The figure for 1999 was £1.8 billion. The total area of agricultural land is 18.6 million hectares, 76 per cent of the entire land surface. According to an agricultural census in June 1999, there has been a decrease of 5.3 per cent in the area given over to crops, as a result of a decrease in cereals and an increase in set-aside. According to a recent report from the Ministry of Agriculture, Fisheries and Food (Maff), the 1999 figures show a drop in the labour force of 3.6 per cent, the largest decrease in a dozen years. 'These results,' the Report continues, 'are not unexpected given the financial pressures experienced by most sectors of the industry over the last few years.'

Farmers' income fell by over 60 per cent between 1995 and 1999. Despite increases in production, earnings were lower in 1999 by £518 million.[2] The value of wheat fell by 6.5 per cent and barley by 5.4. Pigs were £99 million down on 1998 and lambs £126 million down; the value of poultry meat fell by £100 million or 7.4 per cent; the value of milk fell by £45 million; and the value of eggs by 10 per cent or £40 million. A giant profit gap has opened up throughout the industry: rape seed, for example, which costs £200 a ton to produce, is selling for £170 per ton (including the Government subsidy); a savoy cabbage, costing 13 pence to produce, is sold by the farmer for 11 pence, and by the supermarkets for 47 pence.

Hill farmers earned less than £8000 a year on average in 1998–99 (and 60 per cent of that came to them in subsidies), but late last year, when I first started talking to farmers, many were making nothing at all, and most were heavily in debt to the bank. A suicide helpline was set up and the Royal College of Psychiatrists expressed concern at the increased number of suicides among hill farmers in particular. A spokesman for Maff said that agriculture was costing every British taxpayer £4 a week. After Germany and France, the UK makes the largest annual contribution to the Common Agricultural Policy, and yet, even before the great rise in the strength of the pound, British farmers' production costs were higher than anywhere else in the EU, to a large extent because of the troubles of recent years.

'Everything is a nightmare,' one farmer told me. 'There are costs everywhere, and even the subsidy is spent long before you receive it. We are all in hock to the banks – and they say we are overmanned, but we don't have anybody here, just us, and children maybe, and an absolute fucking nightmare from top to bottom.' The strong pound, the payment of subsidy cheques in euros, the BSE crisis, swine fever, and now foot and mouth disease, together with overproduction in the rest of the world's markets – these are the reasons for the worsened situation. But they are not the cause of the longer-term crisis in British farming: local overproduction is behind that, and it is behind the destruction of the countryside too. For all the savage reductions of recent times, farming still employs too many and produces too much: even before the end of February, when diseased livestock burned on funeral pyres 130 feet high, some farmers were killing their own livestock for want of a profit, or to save the fuel costs incurred in taking them to market.

In Britain nowadays most farmers are given aid – a great deal of aid, but too little to save them – in order to produce food nobody wants to buy.[3] The way livestock subsidies work – per animal – means that there is an incentive for farmers to increase flocks and herds rather than improve the marketing of what they've got. As things are, subsidies save some farmers, but they are a useless way to shore up an ailing industry, except perhaps in wartime.

The evidence of what is wrong is out in the British land itself. It is to be found in the particularities of farming experience now, but also in a historical understanding of what farming has meant in this country. Farming – more even than coal, more than ships, steel, or Posh and Becks – is at the centre of who British people think they are. It has a heady, long-standing, romantic and sworn place in the cultural imagination: the death of farming will not be an easy one in the green and pleasant land. Even shiny, new, millennial economic crises have to call the past into question. How did we come to this?

In the 18th century, farmers were still struggling out of the old ways depicted in *Piers Plowman*, or the Bayeux Tapestry, where English farm horses are seen for the first time, bringing vegetables from the fields to the kitchen table. Jethro Tull, one of the fathers of modern agriculture, devoted himself to finding ways to increase yields – he invented the seed-drill, a machine that could sow three rows of seed simultaneously – and collected his ideas in *The New Horse Houghing Husbandry: or, an Essay on the Principles of Tillage and Vegetation* (1731). His ideas were widely accepted by the time he died at Prosperous Farm, near Hungerford in Berkshire, in 1741. Arthur Young, an agricultural educator and zealot of Improvement, set out in 1767 on a series of journeys through the country. *A Six Months' Tour through the North of England* gives a spirited first-person account of changing agricultural conditions. 'Agriculture is the grand product that supports the people,' he wrote, 'both public and private wealth can only arise from three sources, agriculture, manufactures and commerce ... Agriculture much exceeds the others; it is even the foundation of the principal branches.' But the new improvements came at a price and they changed for ever the relationship between the land and the people who tried to live by it. British peasant life was effectively over. 'The agrarian revolution was economically justifiable,' Pauline Gregg writes in *A Social and Economic History of Britain 1760–1965*, but 'its social effects were disastrous. Scores of thousands of peasants suffered complete ruin. The small farmer, the cottager, the squatter, were driven off the soil, and their cottages were often pulled down.' The British countryside, in the face of all improvements, and with every prospect of sharing in the coming wealth of nations, became as Goldsmith described it in' The Deserted Village:'

> Ill fares the land, to hastening ills a prey,
> Where wealth accumulates, and men decay;
> Princes and lords may flourish, or may fade;
> A breath can make them, as a breath has
> made;

But a bold peasantry, their country's pride,
When once destroyed, can never be supplied.

In the spring of 1770 British cows were so disabled by starvation that they had to be carried out to the pastures. This business was known as 'the lifting'. *The General View of Ayrshire*, published in 1840, records that as late as 1800 one third of the cows and horses in the county were killed for want of fodder. By the end of winter in this period, according to John Higgs's *The Land* (1964), every blade of grass had been eaten and the animals were forced to follow the plough looking for upturned roots.

The social structure of the country had changed, the population had grown, the plough had been improved, the threshing machine had been invented, and crop rotation had taken hold. William Cobbett, in his *Rural Rides* – originally a column that appeared in the *Political Register* between 1822 and 1826 – captured the movements which created the basis of the farming world we know. Cobbett rode out on horseback to look at farms to the south and east of a line between Norwich and Hereford; he made an inspection of the land and spoke to the people working on it. He addressed groups of farmers on the Corn Laws, taxes, placemen, money for agricultural paupers, and the general need for reform.

In one of his columns he describes meeting a man coming home from the fields. 'I asked him how he got on,' he writes. 'He said, very badly. I asked him what was the cause of it. He said the hard times. "What times?" said I; was there ever a finer summer, a finer harvest, and is there not an old wheat-rick in every farmyard? "Ah!" said he, "they make it hard for poor people, for all that." "They," said I, "who is they?"' Cobbett yearned for a pre-industrial England of fine summer days and wheat-ricks, and yet his conservatism did not prevent him from becoming an evangelist of Improvement. As for 'they' – Cobbett knew what was meant; he later called it 'the Thing', and sometimes 'the system'. He railed against everything that was wrong with English agriculture: low wages, absentee landlords, greedy clergymen, corruption; and he was prosecuted for supporting a riot by these same agricultural workers the year after he published *Rural Rides*. Cobbett saw how self-inflated governments could sit by and watch lives crumble. His discriminating rage has the tang of today. 'The system of managing the affairs of the nation,' he wrote in *Cottage Economy*, 'has made all flashy and false, and has put all things out of their place. Pomposity, bombast, hyperbole, redundancy and obscurity, both in speaking and writing; mock-delicacy in manners, mock-liberality, mock-humanity ... all have arisen, grown, branched out, bloomed and borne together; and we are now beginning to taste of their fruit.'

Rain was running down Nelson's Column and Trafalgar Square was awash with visitors inspecting the lions. An American woman stepping into the National Gallery was worried about her camera lens. 'This British weather will be the end of us,' she said, as her husband shook out the umbrellas. In the Sackler Room – Room 34 – children with identical haircuts sat down on the wooden floor; they stared at the British weather of long ago, spread in oils with palette knives, and they, too, asked why it was always so fuzzy and so cloudy. One group sat around Turner's *Rain, Steam and Speed – The Great Western Railway*. The instructor encouraged them to express something about the atmosphere of the picture. 'Does it make you shiver?' she said. 'It's like outside,' one of the children replied. But most of them were interested in the hare running ahead of the train. 'Will it die?' one of them asked. 'Where is it running to?'

The future. You feel the force of change in some of these weathery British pictures. Over the last few months I kept coming back to this room, and sitting here, further up from the Turners, looking at Constable's *The Cornfield*. We see an English country lane at harvest time where nothing is unusual but everything is spectacular. Corn spills down an embankment, going to grass and ferns, going to pepper saxifrage or hog's fennel, dandelion and corn poppy, down to a stream. Giant trees reach up to the dark, gathering clouds. At their foot, a small boy lies flat on his front drinking from the stream. He wears a red waistcoat and has a tear in the left leg of his trousers. A dog with a marked shadow looks up and past him with its pink tongue out. The sheep in front of the dog are making for a broken gate that opens onto the cornfield. A plough is stowed in a ditch; the farmer advances from the field; and in the distance, which stretches for miles, you see people already at work.

The picture has philosophical currency: people will still say it is an important part of what is meant by the term 'British' – or at any rate 'English'. This is the country delegates sing about at the party conferences, the one depicted in heritage brochures and on biscuit tins, the corner that lives in the sentiments of war poetry, an image at the heart of Britain's view of itself. But here's the shock: it no longer exists. Everything in Constable's picture is a small ghost still haunting the national consciousness. The corn poppy has pretty much gone and so have the workers. The days of children drinking from streams are over too. And the livestock? We will come to that. Let me just say that a number of the farmers I spoke to in the winter of 2000 were poisoning their own fields. The Constable picture fades into a new world of intensive industrial farming and environmental blight.

The Cornfield is said to show the path along which Constable walked from East Bergholt across the River Stour and the fields to his school at Dedham. Last October I made my way to Dedham. It was another wet day, and many of the trucks and lorries splashing up water on the M25 were heading to the coast to join a fuel blockade. On the radio a newscaster described what was happening: 'The situation for the modern British farmer has probably never been so dire, and a further rise in the price of fuel could kill many of them off.'

Before leaving I had rung a pig farmer, David Barker, whose farm is north of Stowmarket in Suffolk. Barker is 50 years old. His family have been farming pigs in Suffolk for four generations; they have lived and worked on the present farm since 1957. He owns 1250 acres and 110 sows, which he breeds and sells at a finishing weight of 95 kilos. Among his crops are winter wheat, winter-sown barley, grass for seed production, some peas for canning, 120 acres of field beans, 30 acres of spring oats and 100 acres of set-aside.

'Five years ago I was selling wheat for £125 a ton and now it's £58.50,' David Barker said. 'I was selling pigs for £90 and now they're down to £65. And meanwhile all our costs have doubled: fuel, stock, fertiliser. There's hardly a farmer in East Anglia who's making a profit. The direct payments from Europe have declined also because they're paid out in euros.'

'What about swine fever?' I asked, innocent of the epidemics to come.

'There are over five hundred farms that haven't been able to move pigs since August,' he said. 'Immediately, this becomes an agricultural nightmare. The pigs are breeding, the feed is extortionate, and you end up relying on things like the Welfare Disposal Scheme, where pigs are removed for next to nothing. Gordon Brown's bright idea: they give you £50 for a pig that costs £80 to produce.'

'What can be done?' The stormy weather was making his phone crackly.

'Well, this Government has no interest in farming,' he said. 'People in the countryside in England feel they are ignored and derided and, frankly, it appears that the Government would be much happier just to import food. This is the worst agricultural crisis in dozens of years. We're not making any money anywhere. Take milk: the dairy farmer receives seven pence in subsidy for every pint; it takes between ten pence and 12 pence to produce and it costs 39 pence when it arrives at your door. A lot of farmers are giving up and many of those who stay are turning to contract farming – increasing their land, making prairies, to make it pay.'

'Is that the only way to reduce costs?'

'Yes, that. Or by going to France.'

David Barker used the word 'nightmare' at least a dozen times during my conversation with him. He told me about a friend of his, another Suffolk farmer, who, earlier in the swine fever debacle, had sold his 250 pigs into the disposal scheme, losing £30 on each one. Barker himself was waiting for results of blood tests to see if his pigs had the fever. 'If it goes on much longer it will ruin me,' he said.

When I arrived at Nigel Rowe's farm near Dedham only the weather was Constable-like. Out of his window the fields were bare and flat. 'European pigmeat is cheaper to produce,' he said, 'because we have higher standards and higher production costs. As soon as foreign bacon gets cheaper by more than ten pence per kilo the housewife swaps. That is the rule.'

I asked him if he felt British supermarkets had been good at supporting bacon produced in Essex or Suffolk. 'The supermarkets have been very clever at playing the different farming sectors off against each other,' he said. 'The Danish model is very centralised – they are allowed to produce and market something called Danish Bacon. We are very regional over here, very dominated by the tradition of the local butcher. Supermarkets want the same produce to be available in Scotland as you get in Sussex. Only the Dutch and the Danish can do that, and some of these foreign producers are so powerful – the Danish producers of bacon are much bigger than Tesco.'

Nigel has 2000 pigs. But he's not making money. As well as working the farm he has a part-time job as caretaker at the local community centre. 'In the 1970s we were all earning a comfortable living,' he said, 'and when I was at primary school in the 1960s at least thirty of my schoolmates were connected with farming. Now, in my children's classes, there are three. I had 120 acres and I had to sell it recently to survive. I also had to sell the farm cottage my mother lived in in order to stay here. That's what I was working on when you came – a little house for my mother.'

He looked out the window at the flatness beyond. 'The arithmetic is simple,' he said. 'When I started in this game it took five tons of grain to buy the year's supply of fuel for the tractor. Now it takes 500 tons. What do you think that means if your acreage is the same? The Government seem hellbent on the old green and pleasant land, but they won't get behind the people who keep it that way.' Nigel sat in his living-room wearing a rugby shirt and jeans speckled with paint from his mother's new house. 'They're not thinking straight,' he said. 'Our product needs to be marketed – branded, with a flag, which is presently not allowed. It's all wrong. We have to import soya as a protein source for our pigs now because we can't use other animal meat or bone fat. But this country imports tons of Dutch and Danish meat fed on bone fat.'

As we walked out of the living-room I noticed there were no pictures on any of the walls. We went outside to the pigsties. The rain was pouring down, the mud thick and sloppy on the ground, and one of Nigel's pigs was burning in an incinerator. As we looked out I asked him what had happened to the land. 'The subsidies from the Common Agricultural Policy have got out of hand,' he said, 'because they are linked to production rather than the environment. Did you know the rivers around here are polluted with fertiliser and crap? We're seeing a massive degradation of rural life in this country. Bakers and dairies have already gone, onions have gone, sugar-beet is gone, beef is pretty much gone, lambs are going.'

Before we went into the sty he asked me if I was 'pig-clean'. 'I'm clean,' I said, 'unless the fever can come through the phone.' Hundreds of healthy-looking pink pigs scuttled around in the hay and the mud. He picked one up. 'Farming is passed down,' he said, 'or it should be. A farm is built up for generation after generation, and when it starts to slip and go – you feel an absolute failure. That's what you feel.'

We went around the farm and Nigel explained how things work. The notebook was getting very wet so I put it away. 'You feel a failure,' he said again, looking into the wind. 'The other night I was at a meeting: 140 farmers at a union meeting paying tribute to four hill farmers under 45 who'd committed suicide.' He leaned against the side of the barn. 'We are no longer an island,' he said, 'everything's a commodity.'

Charles Grey, the leader of the Whig Party, won a snap election in 1831 with a single slogan: 'The Bill, the whole Bill and nothing but the Bill.' The Reform Act, which was passed the following year after several re-versals and much trouble from the Lords, increased the British electorate by 57 per cent and paved the way for the Poor Law and the Municipal Corporations Act; this in turn killed off the oligarchies which had tradi-tionally dominated local government. The misery and squalor that Cobbett had described in the late 1820s worsened during the Hungry Forties; it was not until after the repeal of the Corn Laws, and the subsequent opening up of trade, that British farmers found a brief golden moment. By the end of 1850 Burns and Wordsworth and Constable were dead, and the coun-tryside they adored was subject to four-crop rotation and drainage. Something had ended. And the Census of 1851 shows you what: for the first time in British history the urban population was greater than the rural. Yet the cult of the landscape continues even now as if nothing had changed.

In 1867 it became illegal to employ women and children in gangs providing cheap labour in the fields. This was a small social improvement at a time when things were starting to get difficult again: corn prices fell; there was an outbreak of cattle plague; cheaper produce arrived from America; refrigeration was invented in 1880 and suddenly ships were coming from Australia loaded with mutton and beef. At a meeting in Aylesbury in September 1879, Benjamin Disraeli, by then Earl of Beaconsfield, spoke on 'The Agricultural Situation', and expressed concern about British farming's ability to compete with foreign territories. 'The strain on the farmers of England has become excessive,' he said. The year before, he claimed, the Opposition had set 'the agricultural labourers against the farmers. Now they are attempting to set the farmers against the landlords. It will never do ... We will not consent to be devoured singly. Alone we have stood together under many trials, and England has recognised that in the influence of the agricultural interest there is the best security for liberty and law.' British farming struggled to compete in the open market until 1910, when the Boards of Agriculture and Fisheries and Food were established and the state became fully involved in supporting it. No one was prepared for what was coming next: squadrons of enemy aeroplanes would darken the fields, and out there, beyond the coast, submarines were about to reintroduce the threat of starvation.

Some parts of East Sussex look like Kansas now. For miles all you see is a great rolling carpet of brown or yellow. There are so few trees; sometimes a patch of meadow peeps out, or a factory appears, leading to a giant rustic-style Safeway and a patch of ring-road leading to a town. But mostly you see the featureless prairie leading nowhere. At Bradford Farm, outside Uckfield, there are five modern-looking barns and a house islanded in bushes. At the end of the garden a Union Jack flaps in the wind at the top of a pole. And on the day I came, there were many plums rotting on the trees.

Michael Fordham, the farm's owner, was driving a new JCB in and out of one of the barns, lifting grain into a waiting lorry. The farm where we stood was 290 acres in all. 'But we run a further 700 acres from here,' Michael said, climbing down from the JCB. Michael looks well, with his brown, weathered face and battered jeans. The lorry roared away, taking the grain off to be dried and processed for bread. 'When I was a boy we used to take the stuff off the combine harvester in sacks,' he said. 'There were about ten men running around carrying sacks. Now it never touches human hand. You couldn't find jobs for ten men now.'

Michael has one man: his father Jim. When I asked him to tell me about his family history a glint appeared in Jim's eye. 'My grandfather George was born in 1860,' he said, 'and he worked as a butcher and a restaurant owner up in London, near St Paul's. My father was Arch Fordham and he started a farm in Berkshire, but my mother, Elsie, who was born in 1901, her family was called Stevenson, and they had worked a farm in Fletching since 1760.' All his family were involved in farming one way or another: his older brother, now dead, owned a farm in Hampshire; his younger brother works for an agricultural machinery business in Vancouver Island; his elder sister is married to a farmer and lives five miles away; the other sister had worked for the Milk Marketing Board in Shropshire. 'She used to work for the RAF,' he said. 'I was too young for the war – I did my bit by farming.' After telling me all this he pointed to an endless field beside us. 'A V-1 crashed over there,' he said. 'I well remember all the dog-fights overhead.'

Later on, at the door of one of the barns, I listened to a conversation between Michael and his father.

Jim: We all used to work together. We had time to have a chat and a laugh.

Michael: You can't hear each other now. The machines. Your only contact with people who are working for you is by mobile phone.

Jim: It's hard for young people to come into it now. When I was young there was always the chance of having a farm. Every farm had some beef, a few pigs, a chicken running about. We had green vegetables. We'd kill a pig once a year.

Michael: Now you've got to be really established from the beginning.

Jim: It was a day's work.

Michael: You live as if you're going to die tomorrow but farm as if you're going to be farming for ever.

Jim: That's an old saying, that. An old local saying.

Michael: My eldest son is 17. He was a great help this summer. But I can't pay him what he can get working at the golf course up the road. With a farm nowadays it's all about management and making things balance.

Jim: During the war they would take anything.

Michael: We got two inches of rain on Thursday and the wheat we hadn't yet harvested lost some of its quality.

Jim: You've got to work all hours.

Michael: Do you know what the average age of the British farmer is now? Fifty-eight.

Jim: Fifty-eight. I'm 70. I used to work 16 hours a day.

Michael: I wouldn't want to be a commuter, though. But I suppose people's aspirations are different. Rain is the bane of my life.

Jim: I thought that was me.

They both have strong Sussex accents and similar faces; they wear the same jeans and matching watches. And the conversation they have, back and forth, is very typical of arable farmers who are doing all right. It is not the banter of pig or cattle farmers, or the absolute depression of those on the hills. The Fordhams of Sussex are hard workers and profit-makers; they have certainly known better times, but they are making do, by acquiring new machinery and renting larger acreages to allow the technology to earn its keep. Under the wintry sun they have grown coldly industrial. They have bowed to an intensive agricultural process that is pounding the countryside and killing other farmers, but they feel there is nothing else for it.

'We're not getting sulphur in the rain any more,' Michael said. 'A lot of the factories that produced sulphur have gone. Some of these crops like sulphur so I'll have to plan ahead for that.' They recently built a new barn at a cost of £60,000. 'It's essential now,' Michael said. 'People like Tesco want to know where everything is stored. It's called Product Assurance.' Inside the barn the grain was piled very high; it smelled of rain. On the other side of the hangar there were bags of fertiliser. As we drank tea in his cottage Michael made a rice pudding and put it in the Aga.

'The old farmhouse cooking,' I joked.

'My wife works,' he said. 'You have to pitch in with everything.'

Michael's father brought his tea over to the table. 'It was the wartime government that gave us our head,' he said. 'They promoted agriculture at any cost. The subsidy thing really started with that. The present trouble has a long history, but the war made us produce and produce.'

I asked Jim if he'd been happy as an English farmer. 'I've had an understanding bank manager,' he said. 'But the Government won't protect English farming. You've just got to face it when things are over.'

Just before evening we drove to one of the detached fields that Michael is farming for a bit of extra profit. We stood in a giant expanse of wheat. A few Scots pines in the distance sparred with electricity pylons. As I crouched at the edge of the field, Michael drove down the incline in the biggest machine I have ever seen: a Claas Dominator 218 combine harvester. It mowed further and further away, cutting, separating, collecting, ejecting rubbish out of a pipe at the side. In five minutes the vast field suddenly seemed too small for the great machine. The blades and the wheels trundled on in the soft wind, and I noticed something eerie in the atmosphere. It took me a while to work it out. There were no birds singing.

Nineteen-fourteen was yet another beginning in British farming. John Higgs argues that the war found agriculture singularly unprepared:

> The area under crops other than grass had fallen by nearly 4.4 million acres since the 1870s ... and the total agricultural area had fallen by half a million acres. When the war began the possible effects of submarine attacks were unknown and there seemed no reason why food should not continue to be imported as before. As a result only the last two of the five harvests were affected by the Food Production Campaign. This came into being early in 1917 with the immediate and urgent task of saving the country from starvation.

This was the start of a British production frenzy, a beginning that would one day propagate an ending. Free trade was cast aside in the interests of survival, and Agricultural Executive Committees were set up in each county to cultivate great swathes of new land, to superintend an increase in production, with guaranteed prices. The Corn Production Act of 1917 promised high prices for wheat and oats for the postwar years and instituted an Agricultural Wages Board to ensure that workers were properly rewarded for gains in productivity. Some farmers objected to having their produce commandeered for the war effort. One of them, C.F. Ryder, wrote a pamphlet entitled *The Decay of Farming*. A Suffolk farmer of his acquaintance, 'without being an enthusiast for the war', was

> quite willing to make any sacrifice for England which may be essential, but, as a dealer in all kinds of livestock, he knows the shocking waste and incompetence with which government business has been conducted, and thinks it grossly unjust that, while hundreds of millions have been wasted, on the one hand, there should be, on the other, an attempt to save a few thousands by depriving the agriculturalist of his legitimate profit.

Despite the words of the non-enthusiast, the war had made things temporarily good for farmers. But the high prices of wartime couldn't be maintained and in 1920 the market collapsed. This was to be the worst slump in British agriculture until the present one. With diminished world markets and too much grain being produced for domestic use, the Corn Production Act was repealed in 1921. British farmers were destitute.

In *A Policy for British Agriculture* (1939), a treatise for the Left Book Club, the former Minister for Agriculture, Lord Addison, tried to explain the devastation that took place during those years.

> Millions of acres of land have passed out of active cultivation and the process is continuing. An increasing extent of good land is reverting to tufts of inferior grass, to

brambles and weeds, and often to the reedy growth that betrays water-logging; multitudes of farms are beset with dilapidated buildings, and a great and rapid diminution is taking place in the number of those who find employment upon them ... Since the beginning of the present century nearly a quarter of a million workers have quietly drifted from the country to the town. There are, however, some people who do not seem to regard this decay of Agriculture with much dismay. They are so obsessed with the worship of cheapness at any cost that they overlook its obvious concomitants in keeping down the standard of wages and purchasing power, and the spread of desolation over their own countryside. Their eyes only seem to be fixed on overseas trade.

There are those who argue that it was this depression – and the sense of betrayal it engendered in farmers between the wars – that led the Government to make such ambitious promises at the start of World War Two. Addison's policy, like many agricultural ideas of the time, was based on a notion of vastly increased production as the ultimate goal. 'Nothing but good,' he wrote, 'would follow from the perfectly attainable result of increasing our home food production by at least half as much again ... a restored countryside is of first-rate importance.'

It was too early in the 20th century – and it is perhaps too early still, at the beginning of the 21st – to see clearly and unequivocally that the two goals stated by Addison are contradictory. The vast increases in production at the start of World War Two, and the guarantees put in place at that time, set the trend for overproduction and food surpluses – and began the process of destruction that continues to threaten the British countryside. The pursuit of abundance has contributed to the creation of a great, rolling emptiness. But in the era of the ration-book, production was the only answer: no one could have been expected to see the mountains on the other side.

Two years before Addison took office Thomas Hardy died, and voices were raised in Westminster Abbey, invoking his own invocation of the Wessex countryside:

> Precisely at this transitional point of its nightly roll into darkness the great and particular glory of the Egdon waste began, and nobody could be said to understand the heath who had not been there at such a time. It could best be felt when it could not clearly be seen, its complete effect lying in this and the succeeding hours before the next dawn.

It was in Addison's time that glinting combine harvesters began to appear in the fields.

You hear the Borderway Mart before you see it. Driving out of Carlisle, beyond the roundabouts and small industrial units, you can hear cattle lowing and dragging their chains, and in the car-park there are trucks full of bleating sheep arriving at a market that doesn't especially want them. Inside you can't breathe for the smell of dung: farmers move around shuffling papers, eating rolls and sausages, drinking coffee from Styrofoam cups. Some of them check advertising boards covered with details of machinery for sale, farm buildings for rent – the day to day evidence of farmers selling up. 'It could be any of us selling our tractor up there,' an old man in a tweed cap muttered at me.

The tannoy crackled into life. 'The sale of five cattle is starting right now in ring number one,' the voice said. A black heifer was padding around the ring, its hoofs slipping in sawdust and shit, and the man in charge of the gate, whose overalls were similarly caked, regularly patted it on the rump to keep it moving. Farmers in wellington boots and green waxed jackets hung their arms over the bars taking notes. One or two looked more like City businessmen. The heifer was 19 months old and weighed 430 kg. The bidding was quick and decisive: the heifer went for 79 pence a kilo.

'If you were here in the prime beef ring six or seven years ago,' the auctioneer said later, 'you would have seen the farmers getting about 120 pence per kilo. That is why so many of the farmers are going out of business. Four years ago, young female sheep would be going for eighty-odd pounds, and today they are averaging thirty.'

Back in his office the auctioneer took out his books covering the last few years. 'Last week, bulls were averaging 82.7 pence per kilo. Three years ago ...' He riffled through the pages. 'Three years ago the average for bulls was 101.5 pence, and in October 1995, before BSE, the same bulls were fetching 134.2 pence. Think about that. There it is. Black and white. You can't argue with those figures. Hellish.' Because of government regulations stemming from the BSE crisis, every cow in Britain now has a passport, which tells where and when it was born, and where it has been since. But despite the seriousness of the attempt to tackle BSE, and the relative healthiness (by EU standards) of the meat on sale here, no one abroad wants British beef any more. People are always sensitive about meat, and they find it difficult to forget bad news – a fact which is already turning the BSE, swine fever and foot and mouth crises into a full-blown catastrophe, especially in respect of European buyers.

At the moment there are only two abattoirs that export beef. The market is dead. (Before 1996 British beef was generally considered the best in Europe.) Current EU proposals seem set to tackle overproduction by making subsidy payments relate to acreage rather than headcount. 'It will result in a reduction in livestock units,' the auctioneer said, 'and if we don't have the units it's hopeless for the whole community.' The auctioneer was defending his own business, as he should, but why should taxpayers pay subsidies to farmers?

The auctioneer's boss is more strident. 'The industry's going under,' he says, 'and you'll notice something: the Common Agricultural Policy works for the rest of the European Union but it's not working for us.'

'In what ways?' I asked.

'Well, first, the animal welfare rules are more strongly applied here than anywhere, and in this, as in so many areas, British farmers find themselves not to be on a level playing field. We are not competing on equal ground. The weakness of the euro, with the subsidy payments coming in euros, is a killer;[4] and as well as that we are still being punished over the BSE crisis. There doesn't seem to be the appreciation in some quarters ...'

'You mean Government?'

'I'm not going to say that ... Confidence has dropped. This is the largest mart in the UK, and what we see is that volumes are holding up, but costs have gone out of control: increased animal welfare, water usage, effluent handling and passports. Today alone we will be handling 1500 passports for cattle.'

The boss dropped his middle finger ominously onto the table. 'Unless there is reinvestment the industry will ... well, you check your crystal ball,' he said. 'It is unbelievably serious in the rural community and I don't think people properly appreciate it. It's no longer a North/South divide we have in this country: it's a rural/urban divide. It's easy to work out: after eighteen months or more of breeding and feeding and labour, a farmer today is walking home with £339.70 for a good heifer. Nothing.' Five months after he said this, many of the farmers weren't even getting that. 'Nothing' was the accurate figure; compensation the only hope.

Climbing to Rakefoot Farm outside Keswick, you see nothing but hills and, in the distance, the lakes like patches of silver; tea shops and heritage centres and Wordsworth's Walks serve as punctuation on the hills going brown in the afternoon. Will Cockbain was sitting in front of a black range in a cottage built in 1504. 'There are ghosts here,' he said, 'but they're mostly quite friendly.'

Will's father bought Rakefoot Farm in 1958, but his family have been working the land around Keswick for hundreds of years. 'There are more Cockbains in the local cemetery than anything else,' he said, 'and they have always been sheep farmers. Sitting here with you now, I can remember the smell of bread coming from that range, years ago, when my grandmother was here.' Will has 1100 Swaledale sheep and 35 suckler cows on the farm. 'Seven thousand pounds is a figure you often hear as an annual earning for full-time farmers round here,' he said. 'Quite a few are on Family Credit – though not many will admit it. We farm 2500 acres, of which we now own just 170, the rest being rented from five different landlords, including the National Trust. The bigger part of our income comes from subsidies we get for environmental work – keeping the stonewalls and fences in order, maintaining stock-proof dykes, burning heather, off-wintering trees.'

'Can't you make anything from the sheep?' I asked.

'No,' he said. 'We are selling livestock way below the cost of production. Subsidies were introduced in 1947 when there was rationing and food shortages, and the subsidies continued, along with guaranteed prices, and now even the subsidies aren't enough. We've got the lowest ewe premium price we've had for years. In hill farming the income is stuck and the environmental grants are stuck too. Fuel prices are crippling us. We are in a job that doesn't pay well and we depend on our vehicles. We are responsible for keeping the landscape the way people say they are proud to have it – but who pays for it? The people down the road selling postcards of the Lake District are making much more than the farmers who keep the land so photogenic.'

Will Cockbain was the same size as the chair he was sitting in. Staring into the fire, he waggled his stocking-soled feet, and blew out his lips. 'I think Margaret Thatcher saw those guaranteed prices farmers were getting and just hated it,' he said, 'and now, though it kills me, we may have to face something: there are too many sheep in the economy. Farmers go down to the market every other week and sell one sheep, and then they give thirty or forty away. They're not worth anything. There are mass sheep graves everywhere now in the United Kingdom.' Will laughed and drank his tea. 'It's only those with an in-built capacity for pain that can stand the farming life nowadays,' he said. 'I like the life, but you can't keep liking it when you're running against the bank, when things are getting out of control in ways you never dreamed of, relationships falling apart, everything.'

On the walls of Will Cockbain's farm there are dozens of rosettes for prize-winning sheep. A picture of Will's son holding a prize ram hangs beside a grandfather clock made by Simpson of Cockermouth, and an old barometer pointing to Rain. 'This is a farming community from way back,' Will said, 'but they're all getting too old now. Young men with trained dogs are a rarity, and hillfarming, of all kinds, needs young legs. We've lost a whole generation to farming. My boys are hanging in there for now, but with, what, £27,000 last year between four men, who could blame them for disappearing?'

Another day, in the tiny kitchen of a hill farm above Bewcastle, eight hundred feet above sea level, a little girl called Louise Carruthers walks to the sink in her blue parka, carrying a jug of water. She smiles and looks like a girl in a Vermeer. 'I've made nothing this year,' her father, Brian, tells me. 'I'm on Family Credit and dependent on Family Allowance to get from week to week. And I'm working from daylight to dark. I'm trying to encourage my boy not to go into this. One of my friends gave up and now he's working on a building site earning £280 a week.'

Brian's farm is a confusion of mist and rotting leaves. When I arrived he slammed the door of his Land-Rover shut and tasted the air. 'Last year we were buying fuel at 11 pence a litre, and now it's 22 pence, and has even been up as high recently as 27 pence.' Inside the cottage, there is a pair of Massey Ferguson overalls hanging above the Aga and a waxcloth on the table. 'You know all the reasons for this,' he said, 'but people ought to look at the supermarkets: they are really taking the piss. They talk about partnerships and all that but they have the farmers at each other's throats. Forget the landscape, forget the culture, and forget traditions, forget the efforts being made by farmers to produce good quality stuff – most supermarkets will buy from wherever they get the best deal. They don't support the British farmer, although it suits them to say they do. They squeeze everyone.'

Couldn't Brian diversify? 'Oh aye,' he said, 'they're always talking about diversify. Into what? It's hard to believe the way people plough nowadays. You can't buy into that now, not with those giant agribusiness people, who buy everything and turn it arable. Tenant farmers are no match for those big businesses. Diversify? How else? It's heavy clay land up here, it's wet, there's no tourism, no ponies, no golf.'

'So what do farmers do who can't cope?' I asked.

'Commit suicide,' he said, 'or drive a dumper.'[5]

Brian is 42 years old. He is separated with two children. His wife left him after beginning an affair with a man she met in the office where she worked. He keeps all his cattle passports in a tin and at one point he held a passport up and said to Louise: 'It looks a bit like our Family Allowance book.' They'd been watching *Who Wants to Be a Millionaire?* the previous night, 'and this guy lost £125,000 on a question we knew the answer to. It was amazing, wasn't it, Louise?'

Out in the yard I saw some of Brian's Galloway

cows. It takes three years for these famous cows to mature but under the new BSE regulations cows have to be slaughtered before their 30th month. Brian patted one of the cows on the way past. 'It's going to be a nightmare if they let farming go the same way as mining,' he said, 'but still, we all vote Conservative up here. The Tories were much better to us.'

You could see the Solway Firth twenty miles away. There it was: Scotland and England balanced along a line of blue mist. And it felt as if there was a freeze coming down. Brian nudged me with his arm as I left, and he laughed, looking into the wild unpeopled hills. 'If you find me a new wife make sure she's an accountant,' he said.

As early as 1935 there was panic in the Ministry of Agriculture about the possibility of another war. The First World War had caught British agriculturalists on the hop: this time preparations had to be made. And it was this panic and this mindfulness that set in train the subsidy-driven production that many feel has ruined (and saved) the traditional farming economy in this country, creating an 'unreal market' and a falsely sustained industry, the root of today's troubles. Before the outbreak of war, policies were introduced which favoured the stockpiling of tractors and fertilisers; there were subsidies for anyone who ploughed up permanent grasslands; agricultural workers were released from war duty; and the Women's Land Army was established. Farming became the Second Front, and the 'Dig for Victory' campaign extended from public parks to private allotments.

With the war at sea British food imports dropped by half while the total area of domestic crops increased by 63 per cent: production of some vegetables, such as potatoes, doubled. Farmers in the 1930s had complained that their efforts to increase production in wartime had not been rewarded by an undertaking of long-term government support. The mistake would not be repeated. Promises were made at the start of the war, and in 1947, with food shortages still in evidence and rationing in place, an Agriculture Act was passed which offered stability and annual price reviews to be monitored by the National Farmers' Union. Parliament instituted a massive programme of capital investment in farm fabric and equipment, and free advice on the use of new technologies and fertiliser was made available. Water supplies and telephone lines were introduced in many previously remote areas. Farmers working the land in the 1950s and 1960s, though there were fewer and fewer of them, had, it's true, never had it so good. At the same time increased use of artificial fertilisers and chemical pesticides meant greater yields and what is now thought of as severe environmental damage: motorway bypasses, electricity pylons, larger fields attended by larger machines, with meadows ploughed up, marshes filled in, woods and grasslands usurped by acreage-hungry crops – what the writer Graham Harvey refers to as 'this once "living tapestry"' was being turned into 'a shroud ... a landscape of the dead'.[6]

Government subsidies and grants in wartime, cemented in postwar policy, prepared British farmers for the lavish benefits they were to enjoy after Britain joined the Common Market in 1973. Today, the Common Agricultural Policy gets a lashing whatever your view of the EU. One side sees quotas and subsidies and guaranteed prices as responsible for overproduction and the creation of a false economy. The other accuses it of being kinder to other European states and not giving enough back to British farmers, a view generally shared by the farmers themselves, but secretly abhorred by the Government, which is handing out subsidies. The two sides agree, however, that the CAP doesn't work, and as I write a new round of reforms is being introduced.

In 1957, when the Common Market came into being, there was a deficit in most agricultural products and considerable variance in priorities from state to state – some to do with climate and dietary needs, some to do with protectionist tendencies. (British farmers who feel ill-served by the CAP often say it was formed too early to suit British needs.) The CAP came into effect in 1964. It was intended to rationalise the chains of supply and demand across member states. This was to be achieved by improving agricultural productivity and promoting technical progress; by maintaining a stable supply of food at regular and sensible prices to consumers; by setting up a common pricing system that would allow farmers in all countries to receive the same returns, fixed above the world market level, for their output. Agricultural Commissioners were given the right to intervene in the market where necessary, and a system of variable levies was established to prevent imported goods undercutting EC production. The vexed issue was the common financing system, which still operates today, and which means that all countries contribute to a central market support fund called the European Agricultural Guidance and Guarantee Fund, or EAGGF. All market support is paid for centrally out of this fund, with budgetary allocations for each commodity sector. Cash is paid out to producers in member states regardless of the level of a country's contribution to the fund.

One consequence of this protectionist jamboree

has been an increase, across the board and in all member states, in the variety and quality of available products, from plum tomatoes and cereals to hams and wines and cheeses, with modern supermarkets now carrying a vastly increased range of produce at comparable prices. (This may have pleased British consumers but it hasn't pleased British farmers, who argue that supermarkets have exploited this abundance, breaking traditional commitments to local producers, and 'shopping around Europe' for supplies which could be got in Britain.) A second consequence has been the familiar overstimulated production and the creation of surpluses. It may even be that by continuing to offer not only guaranteed prices but production subsidies to boot, the CAP can be considered one of the chief instigators of the current crisis.

In the early 1990s, European agriculturalists, seeing the need for the CAP to give direct support to an ailing industry – 'to protect the family farm', as they often put it – and to save the environment, began to speak a different language. The European Commission, in its own words,

> recognised that radical reforms were necessary in order to redress the problems of ever-increasing expenditure and declining farm incomes, the build up and cost of storing surplus food stocks and damage to the environment caused by intensive farming methods. A further

factor was the tensions which the Community's farm support policy caused in terms of the EC's external trade relations. Various measures have been adopted since the mid-1980s to address these problems, e.g. set-aside, production and expenditure quotas on certain products and co-responsibility levies on others. However, these proved inadequate to control the expansion of support expenditure.

A constant refrain during the Thatcher period was that measures like these would only serve to impede market forces. The bucking of market forces, however, was one of the founding principles of the CAP, and even today, when we finally see the bottom falling out of the system of rewards and grants for overproduction, the tendency is towards 'relief' packages, which New Labour support through gritted teeth. It would appear that for a long time now British farming has been faced with two choices: a slow death or a quick one. And not even Thatcher could tolerate a quick one.

The first round of CAP reforms, in December 1993, had three main results: better prices for consumers, a scaling down of production and concomitant reduction in surpluses, and greater attention to matters of environmental concern. The benefit to British farmers took the usual form: subsidies. Burning heather and building fences brings in a few extra euros. But the cost to the state remains unbelievably high: of just over £600 million in aggregate farming income in Scotland in 1995, £400 million came in financial aid.

The EU's *Agenda 2000* set out further CAP changes for the new millennium, made necessary in part by the potential eastward expansion of the Union ('major income differences and other social distortions' would arise, and surpluses would grow even larger). The Commission proposed to cut the beef price guarantee by 30 per cent between 2000 and 2002, and to compensate any loss of income with direct payments; member states would be given a degree of independence in deciding how the money was spent. A similar approach was put forward for the dairy sector: the present quota system will remain in place until 2006, by which time average support prices are to be cut by 10 per cent; this will be offset by a yearly payment for dairy cows. A 20 per cent cut in the cereals intervention price in 2000, accompanied by flat-rate partial compensation for arable crops and the abolition of compulsory set-aside, is meant to prevent a heavy potential increase in cereals surpluses, which could reach 58 million tons by 2005. There is to be greater emphasis on 'a more environmentally sensitive agriculture'. Subsidies higher than 100,000 euros per farm are to be capped.

Consumers stand to save more than a billion

pounds from the cuts in support prices; and the Blair Government is largely in agreement with these proposals, although there are elements which, according to documents available from the Scottish Executive,[7] it finds less than satisfactory:

> While the general proposals for addressing rural policy lack detail, they look innovative and offer possibilities for directing support to rural areas. The downside of the proposals is that the compensation payments look to be too generous, there is no proposal to make farm payments degressive or decoupled from production ... The Government has also declared its opposition to an EU-wide ceiling on the amount of direct payments which an individual producer can receive. Because of the UK's large average farm size, this proposal would hit the UK disproportionately. Elsewhere, there is uncertainty about how the proposals would work in practice. This includes the proposal to create 'national envelopes' in the beef and dairy regimes within which Member States would have a certain discretion on targeting subsidies.

A modern journey across rural Britain doesn't begin and end with the Common Agricultural Policy. Since the end of the Second World War, and escalating through the period since the formation of the EEC, what we understand as the traditional British landscape has been vanishing before our eyes. Something like 150,000 miles of hedgerow have been lost since subsidies began. Since the underwriting of food production regardless of demand, 97 per cent of English meadowlands have disappeared. There has been a loss of ponds, wetlands, bogs, scrub, flora and fauna – never a dragonfly to be seen, the number of tree sparrows reduced by 89 per cent, of song thrushes by 73 per cent and of skylarks by 58 per cent. 'Only 20 acres of limestone meadow remain in the whole of Northamptonshire,' Graham Harvey reports. 'In Ayrshire only 0.001 per cent remains in meadowland ... None of this would have happened without subsidies. Without taxpayers, farm prices would have slipped as production exceeded market demand ... Despite years of overproduction, farmers continue to be paid as if their products were in short supply.'

I set out on my own rural ride feeling sorry for the farmers. I thought they were getting a raw deal: economic forces were against them, they were victims of historical realities beyond their control, and of some horrendous bad luck. They seemed to me, as the miners had once seemed, to be trying to hold onto something worth having, a decent working life, an earning, a rich British culture, and I went into their kitchens with a sense of sorrow. And that is still the case: there is no pleasure to be had from watching farmers work from six until six in all weathers for nothing more rewarding than Income Support. You couldn't not feel for them. But as the months passed I could also see the sense in the opposing argument: many of the bigger farmers had exploited the subsidy system, they had done well with bumper cheques from Brussels in the 1980s, they had destroyed the land to get the cheques, and they had done nothing to fend off ruin. When I told people I was spending time with farmers, they'd say: how can you stand it, they just complain all day, and they've always got their hand out. I didn't want to believe that, and, after talking to the farmers I've written about here, I still don't believe it. But there would be no point in opting for an easy lament on the farmers' behalf, despite all the anguish they have recently suffered: it would be like singing a sad song for the 1980s men-in-red-braces, who had a similar love of Thatcher, and who did well then, but who are now reaping the rewards of bad management. As a piece of human business, British farming is a heady mixture of the terrible and the inevitable, the hopeless and the culpable, and no less grave for all that.

Britain is not a peasant culture. It has not been that for over two hundred years. Though we have a cultural resistance to the fact, we are an industrial nation – or, better, a post-industrial one – and part of the agricultural horror we now face has its origins in the readiness with which we industrialised the farming process. We did the thing that peasant nations such as France did not do: we turned the landscape into a prairie, trounced our own ecosystem, and with public money too, and turned some of the biggest farms in Europe into giant, fertiliser-gobbling, pesticide-spraying, manufactured-seed-using monocultures geared only for massive profits and the accrual of EU subsidies. A Civil Service source reminded me that even the BSE crisis has a connection to intensive agribusiness: 'feeding animals with the crushed fat and spinal cord of other animals is a form of cheap, industrial, cost-effective management,' he said, 'and it would never have happened on a traditional British farm. It is part of the newer, EU-driven, ultra-profi-teering way of farming. And look at the results.'[8] Farms in other parts of Europe, the smaller ones dot-ted across the Continent, have been much less in-clined to debase farming practices in order to reap the rewards of intensification.

The way ahead is ominous. In a very straightforward sense, in the world at large, GM crops are corrupting the relation of people to the land they live on. Farmers were once concerned with the protection of the broad biodiversity of their fields, but the new methods, espe-cially GM, put land-use and food production into the hands of corporations, who are absent from the scene and environmentally careless. By claiming exclusive intellectual property rights to plant breeding, the giant seed companies are gutting entire ecosystems for straight profit. In *Brave New Seeds: The Threat of GM Crops to Farmers*, Robert Ali Brac de la Perriere and Franck Seuret investigate the hidden effects of GM-promoted intensification.[9] 'Today,' they write,

> the biotechnological giant Monsanto and others claim a monopoly on plant breeding, not directly but through patents and transgenic seeds. They sell very expensive seeds which require heavy agricultural inputs. Worse, the farmer has no right to reuse them for further sowing or cross-breeding ... It is the genetic revolution that may engender a new way of intensified agriculture and the programmed elimination of small farmers.

It is happening in India, Algeria, and increasingly in places like Zimbabwe,[10] and it is among the factors threatening to make life hell for the traditional farmers of Yorkshire and Wiltshire.

In 1998, in a leaked document, a Monsanto researcher expressed great concern about the unpopularity of GM foods with the British public, but was pleased to report that some headway had been made in convincing MPs of their potential benefits. MPs and civil servants, the document says, have little doubt that over the long term things will work out, with a typical comment being: 'I'm sure in five years' time everybody will be happily eating genetically modified apples, plums, peaches and peas.'

In 1999 the Blair Government spent £52 million on developing GM crops and £13 million on improving the profile of the Biotech industry. In the same year it spent only £1.7 million on promoting organic farming. Blair himself has careered from one end of the debate to the other, swithering between his love of big business and his fear of the *Daily Mail*. Initially, he was in favour of GM research in all its forms: 'The human genome is now freely available on the Internet,' he said to the European Bioscience Conference in 2000, 'but the entrepreneurial incentive provided by the patenting system has been preserved.' Other voices – grand ones – disagreed. 'We should not be meddling with the building blocks of life in this way', Prince Charles was quoted on his website as saying.[11] The Government asked for the remarks to be removed. 'Once the GM genie is out of the bottle,' Sir William Asscher, the chairman of the BMA's Board of Science and Education remarked, 'the impact on the environment is likely to be irreversible.' The Church of England's Ethical Investment Advisory Group turned down a request from the Ministry of Agriculture to lease some of the Church's land – it owns

125,000 acres – for GM testing. More recently, Blair has proclaimed in the *Independent on Sunday* that the potential benefits of GM technology are considerable, but he has also introduced the idea that his Government is not a blind and unquestioning supporter. 'We are neither for nor against,' said Mo Mowlam.

Poorly paid, unsung, depressed husbanders of the British landscape, keeping a few animals for auld lang syne, and killing the ones they can't afford to sell, small farmers like Brian Carruthers, the man who lives outside Keswick with his Galloway cows and keeps his children on Family Credit, or the pig farmers in Suffolk, told me they felt as if they were under sentence of death from the big agricultural businesses. I asked one of them what he planned to do. His response was one I had heard before. 'Move to France,' he said, with a shrug. Graham Harvey is in no doubt about where the fault lies: 'In the early 1950s,' he writes, 'there were about 454,000 farms in the UK. Now there are half that number, and of these just 23,000 produce half of all the food we grow. In a period of unprecedented public support for agriculture almost a quarter of a million farms have gone out of business ... It is the manufacturers and City investors who now dictate the UK diet.'

The Government has been stuck in farming crisis after farming crisis, but it recognises – though until now somewhat mutedly – the accumulating evils of the subsidy-driven culture. In a White Paper introduced in Parliament at the end of last year,[12] the New Labour view of agricultural progress was clarified:

> Subsidies which simply reward production have damaged the countryside and stifled innovation. A complicated bureaucracy has created expensive surpluses of basic products and has prevented farmers from responding to what customers really want. The CAP must be further deregulated so that agricultural production can adapt to a competitive world market. Production quotas which prevent farmers from responding to the market must be removed ... In a few years we will have an expanded EU with up to 12 more member states and a total population of 500 million people. Without CAP reform, the budgetary consequences would be unsustainable. Negotiations have also begun on liberalising agricultural trade in the World Trade Organisation. This will open up new markets as well as exposing us to greater competition ... The Government recently secured EU approval for the England Rural Development Programme, which includes a major switch of CAP funds from production aids to support for the broader rural economy. We will spend £1.6 billion by 2006 – around 10 per cent of total support for the agriculture industry – on measures to advance environmentally beneficial farming practices as well as on new measures to develop and promote rural enterprise and diversification, and better training and marketing.

This is the Government's public position: large-scale, environmentally-friendly tinkering with European funding, attended by vague worries about changes in the world market. An unofficial spokeswoman for Maff told me there were much deeper worries than the policy-wonks would be heard admitting to. 'It is like the end of the British coal industry,' she said:

> but no one wants to be Ian McGregor. In the time since BSE 110,000 head of cattle have disappeared: it seems that farmers were burning them on their own land. It's a cultural thing, too: no one wants to admit that a certain kind of farming, a certain way of English life, has now run to the end of the road. People will supposedly always need bread. But there is no reason to believe it will have to be made with British ingredients. The disasters in farming aren't so temporary. And they aren't mainly the result of bad luck. No. Something is finished for traditional farming in this country. Not everything, by any means, but something – something in the business of British agriculture is over for good, and no one can quite face it.

The day before I set off for Devon there was a not entirely encouraging headline on the front of the London *Evening Standard*: 'Stay Out of the Countryside'. Just when it seemed there was little room for disimprovement in the predicament of British farmers, news came of the biggest outbreak of foot and mouth disease in more than thirty years. Twenty-seven infected pigs were found at Cheale Meats, an abattoir in Essex, a place not far from Nigel Rowe's pig farm in Constable country. Infected animals were quickly discovered on several other farms. Suspicious livestock began to be slaughtered in their hundreds. Such was the smoke from the incineration site in Northumberland that the A69 had to be closed for a time. British exports of meat and livestock (annual export value £600 million) as well as milk (of which 400,000 tons are exported a year) were banned by the British Government and the EU. 'It is like staring into the abyss,' Ben Gill, the President of the National Farmers' Union, said. 'On top of the problems we have had to face in the last few years, the impact is unthinkable.'

The National Pig Association estimates that the relatively small outbreak of swine fever last year cost the industry £100 million. The last epidemic of FMD, which took hold in October 1967, led to the slaughter of 442,000 animals – a loss of hundreds of millions of pounds in today's terms, only a fraction of which made it back to the farmer in compensation. This year's ban has affected more than half of Britain's farmers and no one doubts that many of them will be ruined.

The county of Devon seemed dark green and paranoid when I travelled there the day after the ban was introduced. It seemed to sit in fear of the disinfecting gloom to come, and as the fields rolled by, I considered the ongoing assault on Hardy's Wessex, the trouble on all sides, and the sense of an ending. Yet I'd originally planned my visit here as an opportunity to gaze at a vision of farming success. Stapleton Farm, my destination, was the one named by the Sainsbury's executives, the day I walked with them around the flagship store on the Cromwell Road, as an example of the new kind of partnership that can exist between supermarkets and farmers. Stapleton produces the quality brands of yoghurt and ice-cream admired by Sainsbury's: their optimism seemed hard to recapture on the way to Devon that morning.

Passing through Bideford – Charles Kingsley's 'little white town' – you get a sudden rush of the way things used to be. Local records tell of how, in the 18th century, a bell would ring in Bideford market at 1 p.m., calling the local people to buy wheat. Traders were not allowed to buy until after 2, to prevent dealers from overcharging the poor. The New Market, which opened in September 1960, sold 2317 animals on its first day – dairy cows, calves, sheep and pigs. According to a recent correspondent in *Devon Life*, Bideford Livestock Market has weathered several outbreaks of foot and mouth disease. Peter Kivell, a local farmer, remembers the phone call in the small hours in October 1967, announcing the first case of FMD in the area: 'I spent the rest of the night wondering how we'd manage.'

'The Devonshire tenant,' according to one local account,

> is at once a dairy farmer, a breeder or feeder of cattle, sheep and pigs, and a grower of corn and cider; and this variety of occupation, arising naturally from the character of the climate and soil of the county, has given him a tone of intelligence and activity which is looked for in vain in other parts of the kingdom, where a monotonous routine narrows the intellect of the dairyman. Farms here are generally of moderate size; for although some farmers hold 700 or 800 acres in several separate farms, the great majority run from 50 or 60 to 200 or 300 acres.

The low green valleys and the high druidical groves of the West Country have been subject to much less of the subsidy-mad prairie-isation you see in other British counties, yet still there has been a vast downsizing of the rural economy. In 1995 78,402 animals were sold in Bideford Market; by 1998 the figure had dropped to 48,826. Many farmers are now running their farms at subsistence levels.

Stapleton Farm is not far from Bideford, nearer Great Torrington, and there isn't a cow to be seen there. They use bought-in milk to make the yoghurt and ice-cream that is so highly regarded by the people at Sainsbury's. No livestock, no fields, no manure, no tractors, just a small manufacturing unit that couldn't be doing better. This is the enterprise Sainsbury's put me onto when I asked about the partnerships with farming that mattered to them. This is the new thing.

I found Carol Duncan in a Portacabin she uses as an office. She was surrounded by Sainsbury's invoices and office stationery. Like her husband Peter, who soon arrived with a marked absence of flat cap or wellington boots, Carol considers herself a modern rural producer. 'I was absolutely delighted when we managed to get rid of the very last cow off this farm,' she said. 'That's the thing about cows, you know, they just poo all the time.' Peter's father and his grandfather had run Stapleton Farm in the traditional West Country way; they had livestock and they worked the fields through thick and thin. 'But from an early age I wasn't interested in that kind of farming,' Peter admitted. 'I wanted to be inside reading books. And then, when my time came, I was interested in the different things you can do with milk. In the 1960s we farmers needed to diversify and head ourselves to somewhere better. The traditional way had been to stand around waiting for the government price review. I wanted to make yoghurt and change things around here. My father would say: "Who's going to milk the cows?"'

'He just wouldn't stop being a farmer, his father,' Carol said.

Peter laughed. 'Yes. But we started with three churns. Carol was an art teacher and that kept us going through the difficult years. We made yoghurt and started selling it to independent schools.'

'That's right,' Carol said. 'If you're paying between £13,000 and £16,000 a year for a school, you want to make sure your children aren't going to be eating rubbish. We had to fight for our markets. In 1994 the price of milk in Devon went up by 29 per cent. We had to increase the price of the yoghurt by 5 per cent and we lost some of our German contracts. I went out and fought to get them back. It was horrible: 200-year-old cheesemakers were shutdown, and hardly a Devon clotted-cream maker was left standing. But there's too much milk. It's in over-supply. Six years ago we thought we were going out of business.'

'We started exporting our stuff,' Peter said, 'to Belgium especially. We supply an upmarket supermarket chain called Delhaize.'

'Until this morning,' Carol said. 'We've just been banned from exporting.'

'We're hoping it will only be a matter of weeks,' Peter said, 'but this is the sort of thing that can ruin people. We're praying it doesn't spread.'

There were a number of people coming and going

outside the Portacabin window. They seemed different from most of the farming people I'd met: they were young, for a start, and they seemed like indoor types, a different colour from the field-workers I'd come across in Essex and Cumbria, Kent and Scotland. The Duncans have over thirty people working at Stapleton Farm – chopping, grating, mixing, packaging, labelling, loading. The buildings where the yoghurt and ice-cream are produced are old farm buildings that have been converted. They look typical enough among the high hedges of North Devon; yet inside each shed there are silver machines and refrigerated rooms that are miles away from the world of cows. Peter tells the story of the Sainsbury's development manager coming down to see them in 1998 as if he were relating a great oral ballad about a local battle or a famous love affair. 'The woman came down,' he said. 'I thought she seemed so fierce. They had already taken samples of our yoghurt away. They said they liked them. But when the woman came that day she just said, "I suppose you'd like to see these," and it was the artwork for the pots. They'd already decided we were going into business. I nearly fell off my chair.'

Carol laughed in recognition. 'Yeah,' she said, 'and they say: "How many of these can you produce a week?" So we started aiming for 10,000 pots a week in a hundred Sainsbury's stores. They were very pleased with the way it was going, weren't they, Peter?'

'Oh yes,' he said, 'and we were putting yoghurt into the pots by hand and pressing the lids on. It was incredibly hard work.'

'Then they wanted to double it,' she said.

'Oh yes,' he echoed, 'they wanted to double it. We had to get better machinery. So it was off to the bank for £80,000. Come February 1999 we were doing 50,000 to 60,000 pots a week.'

Carol swivelled in her office chair. 'We think Sainsbury's are geniuses,' she said. 'We just give them yoghurt and they sell it.'

Stapleton Farm processes all its own fruit by hand. All the milk they use comes from three local farms. Recently, they started giving the milk farmers half a penny more per litre, because of the hard time the farmers are having.

'It's been a music hall joke for years,' Peter said, 'about farmers complaining. But now that the worst has come true the whole thing's beyond belief.'

In the face of all this seriousness, I remembered some lines of George Crabbe's, from 'The Parish Register' (1807):

Our farmers round, well pleased with constant gain,
Like other farmers, flourish and complain.

'The ladies who work for us all come from within three miles of here,' Carol said, 'and they're working for housekeeping money. The farms they live on are struggling and they are here to earn money to feed their kids. But it's a struggle for us too. Most of the people who work here take more away from it than we do, but it's our little dream.'

The Duncans' dream has been one of survival and self-sufficiency, and of being free of that last cow. But as environmentalists they may have trouble living with the price of their own success: expansion. The week I spoke to them they were reeling from having bought a £68,000 machine that wasn't yet working. Sainsbury's want them to produce more and more and they are aware of the fact that doing well entails spending more, so that demand can be met. They are now heavily in debt but also rejoicing at their own success. In the autumn of 1999 their contact at Sainsbury's suggested they have a go at making ice-cream.

'Oh God,' Carol said, blushing at the recollection, 'I didn't know how to make ice-cream. I just made a litre in my little kitchen Gelati and we sent it off. They said they had 80 samples to try. And they decided they liked ours the best. So that was it.'

'Yes,' Peter said, 'that was a visit to the bank for another hundred grand. We had about ten weeks to get the production into full swing. And in the first 12 months of production we sold £750,000 worth of ice-cream.'

Carol is more forthright and I would say more conservative than her husband. She obviously hates the idea of farming but likes the idea of country-related things: 'An art student wouldn't be seen dead near a farm,' she said at one point. 'Farmers just have the wrong attitude.'

'No,' Peter said, 'not all of them. The problem was the Marketing Boards, which gave farmers the wrong idea. They thought someone would just take their produce away and turn it into money. This has been the situation since the end of the war. No other country in Europe was like that. That is why we are so far behind.'

Carol heaved a huge sigh. 'I'm so pissed off about the foot and mouth disease. We had a whole lot of ice-cream going into Spain next week. Not now. I hope it doesn't spread to here.'

'Starting to do business with Sainsbury's feels a bit like being mown down by a bus,' Peter said.

'Yes,' said Carol, 'but I was so relieved when we got rid of that last cow and that old farm. That's the thing with a lot of the farmers around here: they have the potential to get into tourism, get into the farm cottages side, caravans and all that.'

Supermarkets want to be able to rely on volume. If

Stapleton Farm's yoghurt continues to grow in popularity – which it will, as part of Sainsbury's Taste the Difference range for the more discerning shopper, costing 45 pence, against the Economy brand's eight pence – then they will have to get bigger. The charm of Stapleton's smallness cannot last; the supermarket culture requires commitment and tolerance of the highest order from producers. 'I remember once thinking,' Peter said, 'that maybe yoghurt would end up being produced by about three factories in Europe. And it may go like that.'

'Our girls,' Carol said, 'have been brought up to believe that Europe is their oyster. And at this moment we are just what Sainsbury's wants.'

I asked the Duncans if they were worried about having all their eggs in one basket. What happens if people get fed up with Devon yoghurt? What happens if Sainsbury's find somewhere cheaper, or somewhere better able to meet the volume required? Or if it falls for the new kid on the block? Carol met my gaze evenly. 'We'll survive,' she said.

Before going into the factory with Peter I had to put on white boots and a white jumpsuit, sterilise my hands, and pull on a hairnet. Peter stopped in the middle of a chilled room, with the sound of clicking going on further along the line, the sound of mass production. 'This was a cattleshed when I was little,' he said. 'I can remember it quite clearly.' We stood beside a pallet of strawberry yoghurts bound for Sainsbury's. It had the special label already attached. I asked him who paid for the Sainsbury's packaging. 'Oh, we do,' he said.

That afternoon Tim Yeo, the shadow agriculture spokesman, said that the Government had responded in chaotic fashion to a chain of farming crises. 'I wish he would shut up and go away,' Nick Brown replied. 'He is trying to make political capital out of a terrible situation.' And when I was barely out of the West Country news broke of another farm where livestock was found to have contracted foot and mouth disease. The farm was in Devon. And the farmer owned 13 other farms.

The most comprehensive guide to British farming performance is provided by Deloitte & Touche's *Farming Results*. 'Despite cutting costs and tightening their belts,' the report for autumn 2000 concludes, 'farmers have suffered the lowest average incomes since our survey began 11 years ago.' Several facts stand out, so unreasonable do they seem, and so shocking. 'In the last five years the net farm income of a 200-hectare family farm has plunged from around £80,000 to just £8000 ... Those farmers who have expanded their operations dramatically in recent times

... cannot sustain profitability in the face of tumbling commodity prices.' 'The bad news,' says Mark Hill, the firm's partner in charge of the Food and Agriculture Group, 'is that we predict small profits becoming losses in the coming year. This is due to a further fall in output prices and yield plus rising costs of £25 per hectare in fuel alone.'

Meanwhile the supermarkets are happy enough. Since 1995 Tesco's pretax profits have grown from £551 million to £842 million. Asda's have increased from £258 million to £410 million. Safeway's were £176 million and are now £341 million. In 1996 Somerfield made £92 million and in 1999 £208.5 million. And Sainsbury's has gone from a 1995 profit of £809 million to a 1999 figure of £888 million. Sainsbury's remains the biggest supermarket chain although Tesco has gained some ground. According to a recent three-volume report from the Competition Commission,[13] the supermarkets – with Tesco commonly said to be the most guilty – share certain practices that are likely to accelerate their own profits at the expense of farmers and producers. This situation has been too little remarked on: despite complaints of a downturn in profitability over the last five years, British supermarkets have enjoyed massive profits at a time when many of the people who produce the goods they sell have been living close to the edge of bankruptcy.

There are three common perceptions about British supermarkets. The first is that the food sold in them is much more expensive than it is in their EU counterparts or in the US. Second, there's thought to be a great disparity between what is called the farm-gate price (what the farmer sells a product for) and the price of the item on the supermarket shelf. According to the Competition Commission, this 'was seen as evidence by some that grocery multiples were profiting from the crisis in the farming industry'. Third, many people believe that the growth of giant out-of-town supermarkets has been a major factor in the shrinking of town centres and the disappearance of small shops. There is some truth in each of these accusations. Food prices are on average 12 to 16 per cent higher in the UK than in France, Germany and the Netherlands. (This trend is exaggerated by the strong pound.) 'In a competitive environment,' the Commission reports, 'we would expect most or all of the impact of various shocks to the farming industry to have fallen on farmers rather than on retailers; but the existence of buying power among some of the main parties has meant that the burden of cost increases in the supply chain has fallen disproportionately heavily on small suppliers such as farmers.' Yet farmers' troubles are seen to be farmers' troubles.

In market-speak the customer is God. Questions

of environmental balance, social welfare, healthy farming culture and common fairness, are not the concern of the Competition Commission: it is interested only in the extent to which the main supermarkets infringe the customer's right to a good price. Between the lines of their assessments, however, a picture emerges of large-scale indifference to the wellbeing of the agriculture industry on the part of some supermarket chains. Again and again evidence is presented which makes clear the extent to which the suppliers are squeezed by the retailers. For example, supermarkets usually require suppliers to agree to 'various non-cost-related payments or discounts, sometimes retrospectively'; impose charges and make changes to contractual agreements 'without adequate notice'; and transfer risks 'unreasonably' to the supplier.

In other words, if a supermarket shows an interest in taking on a given product for its shelves, it will, by several routes, require payments from the producer in return for the favour. The overall UK market for groceries was worth around £90 billion in 1998. Between them, Asda, Morrisons, Safeway, Sainsbury's, Somerfield and Tesco account for 84 per cent of the total. All the main supermarkets admit to 'requesting payments from suppliers as a condition of stocking and displaying their products'. All of them admitted requesting suppliers 'to make a payment for better positioning of products' in the stores. Many suppliers are familiar with what is known as the 'pay to play' system, which requires them to pay fees to the supermarket chains for the privilege of having their products promoted within the store. This pressure on producers is brought to bear in a number of different ways. The supermarket will do so by:

- Requiring improvements in terms in return for increasing the range and depth of the supplier's products in the supermarket.
- Suggesting to a supplier that it would de-list a product and later withdrawing the suggestion after receiving a discount.
- Requiring suppliers to make payments for a specific promotion.
- Selling products on which the labelling indicates, or might be taken to indicate, that the products were of British origin when they originated overseas. Most of the complaints about this practice came from suppliers of pig meat (pork, ham and bacon). Safeway, Sainsbury's and Tesco did not admit that they engaged in the practice.[14]
- Imposing an unfair balance of risk. This often takes the form of the supermarket seeking retrospective discounts from the supplier which reduce the price of the product agreed at the time of sale; requesting compensation from a supplier when the profits from a product were less than the supermarket expected;

requiring the producer to make payments to cover produce wasted; requiring producers to buy back unsold produce; and failing to compensate the producer when the supermarket chain made an error in planning or forecasting sales.

- De-listing producers/growers who are unable to deliver agreed quantities owing to weather conditions.
- Requiring the producer/grower to bear the cost of surplus special packaging ordered by the supermarket chain for a promotion when sales did not meet expectations.
- Requiring prospective suppliers to contribute to the cost of buyer visits, artwork and packaging design, consumer panels, market research, or to provide hospitality to the supermarket employees.
- Requesting suppliers to contribute specifically to the costs of store refurbishment or the opening of a new store.
- Requiring suppliers to make a financial contribution to the cost of barcode changes or reduced price-marked packs.[15]
- Requiring suppliers to purchase goods or services from designated companies, e.g. hauliers, packaging companies, labellers. One supplier reports being effectively forced to use a certain road haulier, and that haulier now being in a position to dictate transport prices and payment terms. The possibility of finding a smaller transport company is denied. The British Printing Industries Federation says that in order to appear on the approved list of printers sent out by a supermarket to producers, a printer must agree to pay a rebate or levy in favour of the particular super-

market chain. The National Farmers' Union told the Competition Commission that 'some multiples specified which packaging manufacturers and hauliers must be used, denying the growers the opportunity to source from the lowest-cost supplier.'

Not all supermarket chains are guilty of all these things, but according to growers, farmers and suppliers, most of the activities are very common. The buying power of the British stores puts them in an unassailable position. Some suppliers, such as Peter and Carol Duncan, speak of a good working relationship with their main buyer. But most of the suppliers I spoke to expressed dismay at the tactics of the multiples. When I asked Tesco's head office if the company believed an unfair burden of costs was being passed back to its suppliers, it refused to comment.

The sky was pink over Galloway the day the massacre started. It was the end of March, the time when lambs are born, when the earth is moist, but the evening I came the cases of foot and mouth disease were close to 1000, and the burning pyres in the Borders had chased away the smells of ordinariness. I came to the edge of a field and stopped: there were no old flints to be found anymore, no coal to be brought up from the ground, and now the sheep had gone too. For miles there was only empty grass, and dark hedges; the shadows of animals, sick and healthy alike, having disappeared into the running water of the River Esk.

A lighted farmhouse outside Lockerbie had a place in my memory. Once upon a time I had seen it in oils in the painting above my parents' phone table. But now it stood as a matter of fact – a house of ruin at the edge of a saddened field – and the farmer and his wife waited for the killers to arrive and exterminate their flock. Earlier that day I had come through the villages of Ayrshire and down though Dumfries and Galloway like someone driving to the funeral of an old familiar. I travelled into the rain with a sense of many deaths. Outside Kelloholm an Asda lorry rumbled past and splashed the pines. One or two sheep, with new lambs, broke the expanse of greenery up to the town of Kirkconnel, but after that, all the way to Dumfries and Lockerbie, there was emptiness.

The woman in the Kirkconnel newsagent could only smile from the nose down. Her eyes were grey. Spread along the counter in front of her were piles of tabloid newspapers with cover-pictures of Liz Hurley and Pamela Anderson half-naked at the Oscars. The woman wore a rose-patterned overall buttoned up to the throat and behind her the wall was a bank of sweetie jars. I asked her for a quarter of wine gums and a bag of Soor Plooms. 'Och,' she said, very Scottish with her West Coast vowels, 'the auld sweeties.' I asked her if she still used the old measurements. 'We're no' supposed tae,' she said. 'A quarter noo. It's a hunner grammes. That's wit ye want, a hunner grammes.'

'What about the foot and mouth?' I asked. 'Is there a lot of worry here?'

'Aye,' she said, 'wur keepin oor fingurs crossed. It's close by.' She said the people who came to the shop were farmers and old people collecting their papers. 'The neebors are awfy worrit,' she said. 'This thing can just wipe oot a whole livin ye know.'

Behind Drumlanrig Castle there were plastic supports on the new-shooting trees. Rain-bearing northeasterly winds were still carrying the disease. There has been so much life in the land hereabouts that it is hard to contemplate the place in a state of abandonment. The sheep and the cows are as famous here as the hills and the poetry. I imagine the place can bear disaster but not lifelessness; in the modern ether of lowland Scotland, in the landscape, in the language, there are strong links between the past and the present, there's a belief in continuity, and no amount of government-speak can allay the fear that something terrible and final is happening here. 'Some losses can't be compensated,'[16] said one of the farmers I met. The morning I came to Dumfriesshire the Government announced that 420,000 animals had already been slaughtered. Another 300,000 were waiting in line for the kill and the numbers would rise still further. People were beginning to say that Maff should have acted much earlier to contain the epidemic.

I stopped at Ellisland Farm, where Robert Burns lived and wrote from 1788 to 1791. There was a sign hanging on the gate:

Foot-and-Mouth Disease
PLEASE KEEP OUT
Animals on these premises are under observation
This notice is displayed under the advice of the Scottish
Ministers

Burns might have laughed at the idea of the Scottish ministers, but he would have lamented the killing of the animals, and no doubt have deplored, in verse, the fact that future generations would gain so little purchase on the horrors of agricultural life. 'My farm does by no means promise to be such a pennyworth as I was taught to expect,' he said in a letter written 213 years ago on these premises. 'It is in the last stage of worn-out poverty, and will take some time before it pay the rent. I might have had cash to supply the deficiencies of these hungry years, but I have a young brother who is supporting my aged mother, another still younger brother and three sisters, on a farm in Ayrshire; and it took all my surplus, over what I thought necessary for my farming capital, to save not only the comfort but the very existence of that fireside family-circle from impending destruction ... and I am determined to stand by my Lease, till resistless Necessity compel me to quit my ground.'

The road to Moffat was closed. Death, disinfectant, and the remaining snow had made the fields unseasonable: any notion of new life was flushed out with fear and newly expensive chemicals. The Village Fayre, a bakery in Lochmaben, asked farmers to be vigilant during 'this sad time' about the dangers of bringing disinfectant near food. The farmers were expected to put off going to the shop and to remove their overgarments if the need of a loaf became too strong. The new McDonald's on the way to Dumfries had no such qualms. The queue for Happy Meals went all the way out of the door.

The environmental artist Richard Long has invented a sort of autogeography; he turns rural walks into texts, and makes something out of the connection between personal experience and the landscape. The work is a lively, elegiac, often beautiful, commentary on the entanglements of personality and time and space, but in its measurement of change it also conjures a very accurate sense of rural estrangement. His work is lonely. The British landscapes he imagines are alive and primary-coloured, but they seem memorialised. They seem remembered. 'My first work made by walking,' Long says, 'was a straight line in a grass field, which was also my own path, going nowhere. I consider my landscape sculptures inhabit the rich territory between two ideological positions, namely that of making monuments or, conversely, of leaving only footprints.'

I thought of Long on the ride through Dumfriesshire. The abandoned fields of late March seemed to belong to him.

SPRING WALK

PRIMROSES AT 3 MILES
FROGSPAWN AT 18 MILES
A CROW NEST-BUILDING AT 29 MILES
A FARMER SOWING AT 34 MILES
LADYBIRDS AT 38 MILES
SQUIRRELS AT 57 MILES
LAMBS AT 62 MILES
STICKY BUDS AT 67 MILES
A TREE PLANTED AT 70 MILES
A BUTTERFLY AT 85 MILES
BLOSSOM AT 104 MILES
DAFFODILS AT 112 MILES

AVON ENGLAND 1991

As darkness came I went to look at some cows up in the hills. Only a few feet away they stood like boulders. The sky was blue at their backs and the field was thick with mud; they exhaled their white clouds of breath into the evening as if nothing was coming or going in their habitat, as if they were at the very centre of worldly calm. One of the calves – brown ears, white face, tagged 20A – stared through me to nowhere with its big dark eyes.

I joined the killers at six in the morning. At Loreburn Hall in Dumfries the Army – commandeered by vets – was deploying the troops in handing out overalls, disinfectant, guns, oils, knives, wellington boots and packed lunches. As we stood in the queue for boots the killers and the labourers who worked with them were milling around with bacon rolls and Styrofoam teas. 'I've come down from Buckie in Aberdeenshire,' said one, 'and I still haven't found out the price here. What is it?'

'Each kill team has two killers and two labourers,' said Davie, the local man beside me, 'and it's £18 an hour for the killers, including for waiting time, and £8 for the labourers. Then it's a pound for every sheep.'

Davie is a butcher by trade but says he can make more money nowadays boning-out in abattoirs. He is one of those popular, entertaining guys, a wedding singer in his spare time, a guitarist, a happy-go-lucky father of four, and someone with more life and colour about him than you might find in a whole division of Argyll & Sutherland Highlanders. 'Fucking hell,' he said, looking at the soldiers doling out the goods. 'I wonder if there's any chance of getting a Land-Rover out of them cunts.'

Davie's killing partner is Tam. 'I could fucking stand about here all fucking day,' he said, 'no bother to me. Eighteen pound an hour and I've not done one sheep yet. And when you count it all up ...' He rolls his eyes in concentration. (All the time he does this – counting money in his head.) 'If you count it up. Hey. We've made dough just standing here eating rolls and talking shite. Caesar, were you out last night?'

Caesar, one of the labourers, looked up and smiled with a whole row of missing top teeth. 'Aye,' he said, 'but I wasn't that bad. Just four or five pints. But <u>he</u> was out of his fucking head.'

'Oh aye,' said Blackie, his comrade, 'fucking steaming. I only went in there for a game of pool. Bladdered I was. Two o'clock I falls in the door and she's giving it earache.'

'Check the sergeant-major,' said Tam. 'He fucking hates the vets. The vets love bossing everybody about. The squaddies go wild man. They hate it.'

'No wonder,' said Davie. 'It's organised chaos in

here and they vets are full of their own importance.'

'Look at them. Haven't a fucking clue,' said Tam.

'Clueless,' said Blackie.

'Daft as fuck,' said Caesar.

Upstairs the vets were taking instruction from the chief vet under a vast Ordnance Survey map of the Dumfries area. The map was covered with pins – it looked as if the whole area was infected. On a blackboard there was a list of the farms where culling was to take place that day. It gave the name of the farm, the farmer, the number of sheep, the officiating vet and the kill team. 'We want it clean and efficient today,' said the chief vet. 'We want to get ahead.'

The farmhouse in Lockerbie had had its lights burning since early that morning. Five hundred sheep had already been herded into a red barn; another 250 would be penned off in the corners of a nearby field. Davie's kill team left the Loreburn Hall with a map they'd been given upstairs. They piled into one car and immediately started arguing about money. 'It doesn't seem right,' said Caesar. 'I'll be working just as hard as you, and we get a tenner less. That's fucked up.'

'Shut it,' said Tam. 'I've got the licence.'

'But we're doing the job together,' said Blackie, 'sweating just the same, and ...'

'Get a grip,' said Davie. 'Just get a fucking grip.

We'll pool all the money, all right? Everything gets put in the middle, and then we split it up evenly, all right? All right Tam? Now Caesar, shut your fucking mouth and let's get on with this. Money and more money, that's all you cunts talk about. Just stop arguing.'

They didn't need the map. The labourers knew where the farm was. The Army Land-Rover was behind them. At the entrance to the farm there was a carpet held in place by two blocks of wood. It was sodden with disinfectant. Davie, Tam, Caesar and Blackie, now wearing rubber waterproofs flown in from Iceland, stood to be hosed from the neck down with disinfectant. But the power-washer broke down when the job was only half done and the vet was losing his patience.

The soldiers struggled with the machine for ages. The vet looked at the Army guys and the Army guys looked back with contempt. 'Fucking Army engineers!' said the vet. 'You can't even fix a power-washer when it's needed.'

'I'm going to fucking lamp him,' said one of the Army guys off to one side.

Davie phoned a nearby slaughterhouse on his mobile and within half an hour they had a new power-washer. It was now a quarter past nine. 'More organised chaos,' said Davie. 'I've got workers desperate for money, self-important vets, Government folk who are egomaniacs, and Army folk who just want to punch everybody.'

The group crossed the carpet into the farm. On the way up the drive the vet turned to the others with a serious expression. 'We're here to do a job,' he said, 'and I don't want any fucking about. No stupid carry-on. This is this guy's livelihood we're destroying here.' The farmer and his wife were standing outside their house. They were holding hands. 'What happens now,' continued the vet, 'is I'm going up to shake hands with the farmer. I'll introduce you one by one. Shake the farmer's hand and then move on up to the barn.'

The two killers and the labourers stepped up to the farmer and shook his hand. The wife stood motionless and the farmer said nothing. He just nodded. The farmer was in his late fifties, with frizzy grey hair and a gold tooth. He looked stricken, and when the killers wandered towards the barn he turned with his wife and went into his house. He didn't appear again.

Caesar and Blackie were younger than the killers. They were also less in tune with the wishes of the vet and the gravity of the situation. They were excited. Inside the barn there were about five hundred sheep bleating and breathing; the sheep chased each other round the barn and climbed over one another. The smell of disinfectant was strong, but so was the smell of rain-dampened sheep. The vet asked to inspect the equipment. The killers produced two Cash guns – captive bolt guns – and showed him the pellets, red for sheep and green for cows, which they would use according to the size of the animals. They also brought out two six-inch boning knives and two sharpening steels. The vet inspected these and then nodded. 'I want it clean,' he said. 'There's no animal to be put into the loader until it has stopped moving. Let's get started.'

The two killers made an arrangement: Davie would shoot and Tam would 'stick them', stabbing behind the neck to cut the spinal cord. The sheep were leaping around. The two killers moved in together and the labourers stood back. Davie grabbed the first sheep and jammed it against the side of the barn. He pushed his knees into its ribs and locked his forearm under its head. Then he tapped the sheep's head with the side of the gun to make it raise up. When it did so he pointed the gun between its ears and fired. The animal dropped and let out a solitary groan. Tam then stepped in and stuck the knife into the back of the neck and cranked it from side to side. Blood poured onto his hands and down the back of the sheep. After this the vet came up and lifted the sheep's head, tapping its eye with a finger to make sure it was dead. The killers then moved on to the next sheep.

After half an hour there were thirty sheep lying in their own blood. Caesar and Blackie began lifting them by the legs into the bucket of a waiting yellow loader. Each time the bucket was full the loader would drive out into the yard and drop the dead sheep into a truck. Meanwhile a second vet was gathering the lambs in a separate pen. While the adult sheep jostled against one another as the killers worked though them, the new lambs stood still in the pen. The vet produced a large syringe and bent down to pick up the first of the fifty lambs. He turned the lamb onto its back and felt for the heart. Then he injected a quantity of anaesthetic. The lamb simply expired. As he held the first dead lamb in his arms the vet looked up, feeling the legs of the animal. 'Good God,' he said, 'this is prize lamb.'

'Some farmers, the bad ones, aren't bothered to see their flock go,' said Davie. 'They're digging the holes before the sheep are dead, looking forward to their compensation.[16] But this farmer must be absolutely gutted, these are beautiful sheep, reared over generations.'

The labourers were stopped once or twice as they grabbed the bloodied legs of the sheep. 'This one's not dead,' said the vet, 'it looked dead but it's still breathing. Take it over to the side of the barn and let it die. I don't want this farmer looking out of his window and seeing half-alive sheep moving in the truck.' The boys moved the sheep as they were told, but most of the animals Davie and Tam left on the hay were dead, motionless and heavy as if they had never been otherwise.

The smell of blood and sweat was everywhere by two o'clock. There were 350 dead sheep and the barn and the yard looked like something out of Breughel's *The Triumph of Death*. One of the vets went down with Tam to the road where the Army had been standing all day; the troops had packed lunches and canisters of soup for the gang. The pair brought back the provisions and everyone gathered in a second, empty barn to eat their lunch. The sandwiches were ham and cheese; the killers lifted them with their bloody hands, and they unscrewed the tops of bottles of mineral water. Caesar rummaged in a bag of corn Nik-Nax with his stained fingers. The farmer's wife came into the barn with a tray of orange cordial. 'Thank you very much, dear,' said the vet. Some of the boys made a slight bow in her direction. 'Thanks.'

Davie and the vet discussed the day. The vet said he was quite happy with the way it was going, but that he preferred the pithin rod as a method of slaughter. (This is a plastic rod that is entered through the hole in the head to scramble the brain and is believed by many to be cleaner.) There had been one bad incident, though: the driver of the loader had missed the truck and dropped a bucketful of dead sheep on the ground. 'Fucking idiot,' said the vet. 'Right outside the farmer's window.'

When the gang returned to the barn after lunch the remaining sheep had squeezed themselves up at the far end. The straw towards the front was saturated with blood. The killers were faster now and they tried to lighten the situation with jokes and stories. Caesar's day-job is in a slaughterhouse cutting the rear-ends out of cows. 'Don't you be cutting the arseholes out of them sheep,' said Davie.

'Fuck off,' said Caesar.

'It's true,' said Davie, making even the vet laugh. 'He's done nothing else since he left school. Cows' arseholes. That's his life. When other boys were young they used to say to their daddies, "Daddy, when I grow up I want to be an astronaut," or "Daddy, I want to be a train driver." Caesar, when he was a wean, he drove through the countryside with his da and he said: "Daddy, them arseholes are mine!" The crowd in the barn laughed quietly but they continued to kill the sheep and lift them into the loader. Eventually all the sheep in the barn were dead and the gang retreated to the yard. In the barn the straw was red and the air was hot and there was no noise.

The last of the sheep had been gathered in the

field. The man organising them had earlier asked the farmer if he would help shepherd the flock, but the farmer said he couldn't. He wanted to stay indoors with the curtains closed. The rain was pouring and the killers' hands were so numb they could barely feel them. They had cuts on their hands and they struggled to pick the spent pellets out of the guns before loading new ones. Up on the hills a mean wind was blowing through the trees and the last snow was sliding down to the stream. Caesar and Blackie put their hands between their knees in the driving rain. 'This is some way to earn a coin,' said Tam.

A few days later Tony Blair stood in Downing Street. The foot and mouth crisis was so severe the election could not take place on 3 May. This was the first time since the war that an election had been put off in the public interest. Ministry of Agriculture officials argued about whether the mass livestock cull was stopping the spread of the disease. But there were even larger questions. When the spreading has stopped where will British farming be? And will the slaughter of livestock in 2001 come to seem merely the most visible aspect of the death of an industry, a death long foretold?

The Lockerbie farmer came into his barn in the evening after the killers had disinfected everything and gone. And standing there, in the silence, he may have remembered the days when his father's farm was splendid enough to be painted in oils.

A very English drizzle was making a blur of Otford the day I visited. Hedges were loaded like wet sponges, the short grass squeaked underfoot, there was mud in the road and mud at the farm gate, with a cold whiteness in the Kent sky that darkened quickly in the afternoon. Ian and Anne Carter were sitting in the drawing-room of their farmhouse. She is a Justice of the Peace, groomed to a fine point of civic order, wearing a blue suit with a poppy pinned to its lapel. She is well-spoken, opening up her world in good clear Southern English, the language of the prep school and the Shipping Forecast, and her generosity seems to go perfectly with the rationale of the teacups. Ian stretched out his long legs like a teenager: he is likably comfortable with everything he knows and everything he doesn't know; he is right as rain and habitually nice. They both shook their heads.

'You need to have 2500 acres to make farming work nowadays,' he said. 'Not so long ago you could have 600 acres and second-hand equipment and send your kids to a good school and holiday in the South of France. That's all gone now.'

'Absolutely,' said Anne, 'there has been pressure from the fertiliser companies to use certain fertilisers. There are too many sheep owing to these awful subsidies. The whole countryside out there has changed almost beyond recognition.' Over the fireplace hangs a Constable painting: a portrait of one of Anne's ancestors. There is something darkly lively about the picture. For a while we all sat and stared into it. 'It's not at all famous,' Anne said, 'all the famous ones are out there being admired.'

A British Legion-type couple came to lunch. 'It's funny the way things go,' the man said, 'when you think of all those British companies that went to the wall. British manufacturing took such a hammering and now you see that a whole way of living and working has disappeared.'

'Do you think the land will eventually be nationalised and given to the National Trust?' his wife asked.

'You mean heritaged?' someone else said.

'Very interesting what's happened to The Archers,' Anne said. 'This year Nigel, who has the big house, Lower Loxley, is involved in some sort of shooting gallery. They didn't have that before. The Archers has become less and less farming and more sex.'

We braved the weather and walked several miles over the fields. Anne spoke about a spiritual connection she felt with the countryside and a hope she retained in the balance of nature. There were milky pools beside the trees, and when I walked with Ian he tried to give an account of why things had gone the way they had, a story of overproduction and subsidy distortion and diseased animals and the threat of bad seeds. It seemed less imposing that the land belonged to the Carters, and much more interesting, in an easy, uncomplicated way, that they belonged to the land thereabouts. They seemed to walk it knowingly. We stopped at the family chapel, dedicated to St Jude, in a building which dates from 1650. A book inside the chapel tells the story of the ownership of this piece of land – the lay ownership from 1066 to 1521, the removal of the house and the farm from a nobleman to one of the wives of Henry VIII. Across from the chapel is a disused cowshed that Anne's father built in 1946. Water dripped from the lintel, and an inscription is carved above. 'To the glory of agriculture,' it says, 'and the working man.'

NOTES

1 According to *Which?* magazine (28 February), this confidence should not be extended to Sainsbury's chickens. In a survey, the magazine found 22 per cent of the store's chickens to 'contain unwelcome bacteria', including salmonella. This was worse than Safeway (21 per cent) and Tesco (6 per cent).

2 The Maff *Report on Wages in Agriculture* (Stationery Office, 40 pp., £11, 29 October 1999, 0 11 243054 6) asserts that employers agreed that wages were too low, especially for casual workers, and suggested that their pay be increased by 5 per cent. As the Report also notes, however, Government inspectors have had a very low success rate in prosecuting employers who pay below the agreed rate. Anecdotal evidence suggests that many growers, especially in the horticultural sector, still pay illegally low wages.

3   Subsidies available for one animal, dependent on the size of herd and its location: a Suckler Cow Premium, £117 a year; Beef Special Premium, £84; Hill Livestock Compensatory Allowance, £69.75.

4   On 11 January this year the European Commission formally approved a scheme by the British Government to pay £34m in compensation to arable farmers to offset the effects of the weakness of the euro against sterling since July 1999. 'I am delighted to say that the Commission has now approved our scheme,' the Agriculture Minister, Nick Brown, commented. 'This is an exceptional response to very difficult trading conditions.' Bizarrely, though, at the start of the foot and mouth epidemic, the Government began referring to this money as 'compensation' for animals slaughtered as a result of the outbreak. In fact, the money was earmarked to compensate farmers for the euro imbalance, and has nothing to do with foot and mouth disease.

5   *Suicide and Stress in Farmers* by Keith Hawton, Sue Simkin, Aslog Malmberg, Joan Fagg and Louise Harriss (Stationery Office, 122pp., £17.50, 11 December 1998, 0 11 322172 x) shows how far from being a joke Carruthers's comment was.

Research for the book was undertaken 'following recognition of the apparent increased risk of suicide in farmers in England and Wales', and what it showed was that 'farmers contributed the largest number of suicides of all the high-risk occupational groups ... the majority ... faced problems connected to work (including financial problems) at the time of their death.' Almost half these problems were classified as 'major' – 'which meant there was an imminent danger of the farm being lost'.

6 *The Killing of the Countryside* is the most convincing account available of the destruction of British rural life by postwar policymakers and corporate agribusiness. It is the source for several of the figures I use here.

7 The Scottish Executive, Department of Agriculture, Food and Fisheries, has published a series of useful documents on these and other matters relating to CAP, available at www.Scotland.gov.uk

8 Not everyone agrees that the feeding of meat and bone-meal is an indirect result of EU intensification alone. Hugh Pennington's letter in the LRB (25 January) points to the practice's popularity in Britain early in the 20th century. It would probably be more accurate to see this kind

of feeding as one promoted, rather, by
agricultural intensification in wartime.

9   Zed Books, 158 pp., £9.99, 11 December 2000,
1 85649 900 6.

10 'A review of cotton production in Zimbabwe,' the
agronomist Fred Zinanga reports in *Brave New
Seeds*, 'shows that 70 per cent of the total crop
produced annually comes from the small-scale
farming sector. As these farmers can barely
afford inputs, the crop is normally grown on
advances from the cotton companies, which are
later deducted at the end of the season. The
question is, how would these small-scale farmers
be able to afford the purchase of transgenic
seeds, especially those with the Bt gene which
have to be purchased annually? Monsanto is
pushing hard to introduce this crop in Zimbabwe
without going through the normal procedures of
testing the technology and studying its economic
compatibility with the local farming system ... It
is therefore suicidal to encourage farmers to
cultivate the supposedly lucrative transgenic
crops, since their seeds are beyond their means.'

11 www.princeofwales.gov.uk

12 *Our Countryside: The Future. A Fair Deal for Rural
England* (Stationery Office, 176 pp., £28,
28 November 2000, 0 10149092 5, www.wildlife-

countryside.detr.gov.uk/ruralwp/cm4909/
index.htm).

13 *Supermarkets: A Report on the Supply of Groceries from
    Multiple Stores in the United Kingdom* (Stationery
    Office, 3 vols, 1256 pp., £80, 10 October 2000,
    0 10 148422 4, www.Competition-
    Commission.org.uk).

14 Sainsbury's, Tesco and Safeway each expressed
    concern at the approach taken by the
    questioners. Safeway had 'reservations about the
    way in which its answers might be presented in
    the report'. Sainsbury's said that 'the yes/no style
    of format requested by the CC had not always
    been appropriate; hence, in a small number of
    cases it had commented on a practice without
    giving a yes/no answer. The response
    summarised company policy in respect of each of
    the individual practices listed. It was not possible
    to check that every policy had been consistently
    and uniformly applied by each of the 350 or more
    buyers, over the last five years.' Tesco found the
    approach of the questionnaire wholly
    unsatisfactory: 'it said that the fundamental basis
    for buyer/supplier relationships was a two-way
    negotiation process through which both supplier
    and buyer strove to achieve their individual
    objectives. The reality was that both parties were

likely to compromise in order to reach mutually acceptable agreements.'

15 Tesco argued that higher costs resulted from promotions instigated by the producers themselves, which required a barcode change; Sainsbury's that payments were negotiable; Somerfield that payment was not always necessary; Safeway, which did not say that it engaged in the practice, said 'that it would not expect to pay for barcode changes on branded products, and on own-label products the packaging was paid for by the supplier so it, too, would fund the changes.'

16 The Government did eventually get themselves sorted out on financial compensation. For breeding ewes the farmer would receive £90; for rams £150; for new season's lambs, £60. For pigs the figures were better: gilts £190, breeding sows £130, and boars £520. For cattle: £1000 for breeding bulls, £900 for breeding heifers, £1100 for breeding cows, and an average of £550 for clean cattle.

THANKS

When I set out to write this essay I had no idea where it would end. I sensed that farming was in deep trouble, but could not have known how soon that trouble would become part of a national crisis, the worst in British agriculture for over a hundred years. Like the best and the worst of essayists – ever a slave to the egotistical business – I found myself initially taken up with the strangeness of the farming world I had known and ignored as a child; by the end of my six-month tour of duty, however, with mountains of livestock burning on the headland, I felt the impulse to simply be a witness to what was happening. A way of life and a way of landscape may be caught here in a state of eclipse, and yet, I fear, there will be even sadder seasons ahead for British farmers before this matter is finished. But there can be no doubting it: this year has been the most bitter of harvests.

My own gatherings, such as they are, relied on several helping hands, chiefly those of my friend and researcher Jane Swan. This piece was first written for the London Review of Books and I would like to thank the paper's editors for their help with it. Also: the National Farmers' Union, London School of Economics, the Ministry of Agriculture, Fisheries and Food, Sainsbury's Ltd, Stapleton Farm in Devon, Will Cockbain of Keswick, Brian Carruthers of Bewcastle, Michael Fordham of Uckfield, and the many other farmers I spent time with in Sussex, Essex, Norfolk, Devon, Cumbria, Dumfriesshire and Kent. I am also grateful to Tim and Sarah Goad, Richard Long, Kate Griffin, and my chief advisor in the town of Lockerbie, whose name, with several others, has been changed to protect the innocence of this year's necessary killers.

Andrew O'Hagan, April 2001

The End of British Farming *was first published in the*
London Review of Books.

The London Review of Books can be ordered through
your local newsagent or taken on subscription.
For subscriptions please call our 24-hour freephone on
0800 018 5803 (within the UK) or 020 7209 1141
or fax 020 7209 1151.